AF847667

# Membranes for Electrochemical Devices

Membrane technology is vital for electrochemical devices used in energy and environmental applications. This resource explores membrane fundamentals, including selection criteria and ion conduction mechanisms, advances in synthesis techniques and nanomaterial integration, and practical applications, along with discussions on scaling challenges. The content addresses both theoretical principles and practical considerations, offering insights into performance enhancement strategies and future developments that could overcome current limitations in membrane technology.

- Offers an in-depth introduction to various electrochemical devices and explains the crucial role of membranes in these technologies
- Discusses different types of membranes used in electrochemical devices, including ion-conducting, proton-conducting, anion-exchange, and composite membranes
- Covers a wide range of membrane fabrication methods, challenges, and advancements in scaling up membrane production
- Analyses key membrane properties and explains how they directly affect the performance of electrochemical devices
- Explores the distinct requirements of each technology and how membranes can be modified to improve performance

This book is aimed at graduate students and researchers in membrane technology, electrochemistry, and materials science.

# Membranes for Electrochemical Devices

## Materials, Fabrication, and Applications

G. Arthanareeswaran, Pei Sean Goh, and
Hemanth Kumar K

CRC Press
Taylor & Francis Group
Boca Raton London New York

CRC Press is an imprint of the
Taylor & Francis Group, an **informa** business

Designed cover image: Shutterstock

First edition published 2027
by CRC Press
2385 NW Executive Center Drive, Suite 320, Boca Raton FL 33431

and by CRC Press
4 Park Square, Milton Park, Abingdon, Oxon, OX14 4RN

*CRC Press is an imprint of Taylor & Francis Group, LLC*

© 2027 G. Arthanareeswaran, Pei Sean Goh, and Hemanth Kumar K

Reasonable efforts have been made to publish reliable data and information, but the author and publisher cannot assume responsibility for the validity of all materials or the consequences of their use. The authors and publishers have attempted to trace the copyright holders of all material reproduced in this publication and apologize to copyright holders if permission to publish in this form has not been obtained. If any copyright material has not been acknowledged please write and let us know so we may rectify in any future reprint.

Except as permitted under U.S. Copyright Law, no part of this book may be reprinted, reproduced, transmitted, or utilized in any form by any electronic, mechanical, or other means, now known or hereafter invented, including photocopying, microfilming, and recording, or in any information storage or retrieval system, without written permission from the publishers.

For permission to photocopy or use material electronically from this work, access www.copyright.com or contact the Copyright Clearance Center, Inc. (CCC), 222 Rosewood Drive, Danvers, MA 01923, 978-750-8400. For works that are not available on CCC please contact mpkbookspermissions@tandf.co.uk

For Product Safety Concerns and Information please contact our EU representative GPSR@taylorandfrancis.com. Taylor & Francis Verlag GmbH, Kaufingerstraße 24, 80331 München, Germany.

*Trademark notice*: Product or corporate names may be trademarks or registered trademarks and are used only for identification and explanation without intent to infringe.

ISBN: 978-1-041-15365-8 (hbk)
ISBN: 978-1-041-15424-2 (pbk)
ISBN: 978-1-003-67936-3 (ebk)

DOI: 10.1201/9781003679363

Typeset in Times
by SPi Technologies India Pvt Ltd (Straive)

# Contents

# About the Authors

**G. Arthanareeswaran** is Professor and Head of the Department of Chemical Engineering at the National Institute of Technology Tiruchirappalli, India. He holds a B.Tech. in Chemical Engineering from Coimbatore Institute of Technology and completed his M.Tech. and Ph.D. at Anna University, India. With a strong academic and research background, his expertise lies in membrane science and technology, particularly in nanocomposite and mixed matrix membranes for water purification, desalination, and energy systems. His work emphasizes the design and fabrication of nanocomposite and mixed matrix membranes to enhance ion selectivity, conductivity, and durability, contributing to sustainable solutions in environmental engineering and industrial separation technologies. He has authored 220 peer-reviewed publications, books, and book chapters in reputed international outlets. He is the recipient of prestigious international awards, including the RAE (UK) Research Exchange Award, the Endeavour Executive Award (Australia), and the Brain Pool Award (South Korea), Fellow of the Indian Membrane Society, and Fellow of the Indian Desalination Association, reflecting his global research excellence and contributions to membrane science and engineering.

**Pei Sean Goh** is Associate Professor in the School of Chemical and Energy Engineering, Faculty of Engineering, Universiti Teknologi Malaysia (UTM). She received her Ph.D. in Gas Engineering at UTM. Pei Sean is also a research fellow of the Advanced Membrane Technology Research Centre (AMTEC), UTM, and the head of the Nanostructured Materials Research Group at UTM. Her research interests focus on the synthesis of a wide range of nanostructured materials and their polymer-based composites for membrane-based separation processes. Her research mainly focuses on the application of carbon-based nanomaterials and polymeric nanocomposite membranes for acidic gas removal, desalination, and wastewater treatment. Pei Sean is the winner of L'Oreal-UNESCO for Women in Science Malaysia 2020 and the Outstanding Young Chemist Award 2025 of the Malaysia Institute of Chemistry.

**Hemanth Kumar K** is pursuing his PhD in the Department of Chemical Engineering at the National Institute of Technology, Tiruchirappalli. He completed his bachelor's degree in chemical engineering from Anna University, India, in 2022. He completed his postgraduate degree in Chemical Engineering from Anna University, India, in 2024. Later, he joined the PhD programme in July 2024 under the guidance of Prof. Dr. G. Arthanareeswaran, Department of Chemical Engineering, National Institute of Technology Tiruchirappalli, India. Currently, he is working in the research area of membranes for energy and environmental applications. He is the recipient of the A.P.J. Abdul Kalam Young Research Fellowship for the year 2025. He has authored several research and review articles in highly reputed international peer-reviewed journals.

# Preface

Membranes serve as the physicochemical interface between electrodes in electrochemical systems, where their structure and composition directly dictate ion transport behaviour, interfacial stability, and overall device performance. Consequently, surface and bulk modification strategies have been widely adopted to tailor membrane properties and significantly enhance key physicochemical parameters such as ionic conductivity, selectivity, mechanical strength, thermal stability, and compatibility with electrode interfaces.

The functionalization of polymers and nanomaterials enables the introduction of specific chemical moieties and bonding groups, thereby enhancing membrane performance in electrochemical systems. The continuous growth of electrochemical energy technologies for clean energy conversion and storage has placed increasing emphasis on the rational design, selection, and optimization of membrane materials.

The increasing demand for high-performance, safe, and sustainable electrochemical energy conversion and storage technologies has set new objectives for membrane science and engineering. In devices such as batteries, supercapacitors, fuel cells, redox flow batteries, and electrolyzers, membranes play a critical role in regulating ion transport, preventing electrical short circuits, and ensuring long-term operational stability. However, challenges include interfacial degradation, limited ionic conductivity, insufficient chemical and thermal stability, and mechanical failure under prolonged cycling remain major concerns. To address these issues, extensive research has focused on the rational design of advanced membrane materials and architectures with tailored ionic conductivity, selectivity, mechanical robustness, and long-term stability. These efforts encompass the optimization of polymer backbones, the introduction of functional groups, precise control over membrane morphology and thickness, and the engineering of stable interfaces with electrodes and electrolytes. In this context, this book provides an overview of key strategies used to synthesize functional polymers and fabricate membranes for electrochemical applications, including surface functionalization, interfacial polymerization, electrospinning, solvothermal deposition, and layer-by-layer self-assembly. State-of-the-art examples are discussed to illustrate how controlled structures and properties can be achieved, while current limitations, practical challenges, and future opportunities in membrane materials, fabrication routes, and device integration are also highlighted, with the overarching goal of enabling durable, safe, and cost-effective next-generation electrochemical energy systems.

The primary audience of this book includes graduate students, researchers, scientists, and professionals working in the areas of electrochemistry, materials science, and membrane technology. This advanced contribution aims to provide an integrated perspective on the selection and design of membrane development in electrochemical devices. Although significant progress has been made, a comprehensive treatment focusing specifically on the role of functional membranes in electrochemical systems remains limited, and this book seeks to address that gap. Today, proton-exchange membranes, anion- and cation-exchange membranes, composite and solid-state membranes constitute the core material platforms for a wide range of electrochemical applications. Their performance is governed by key parameters such as ionic conductivity, permeability, selectivity, mechanical strength, chemical and thermal stability, all of which directly influence device efficiency, durability, and safety.

Therefore, the development of novel functional polymers and hybrid membrane architectures is essential to further enhance electrochemical performance. Furthermore, self-standing membranes with tuneable pore size, surface functionality, and defect-free structures show great promise for next-generation electrochemical devices. This book is organized into two main sections. The first section defines, classifies, and discusses the fabrication and properties of membranes for electrochemical applications, including material selection, synthesis routes, and modification strategies. The second section highlights the applications of membranes in electrochemical devices, emphasizing their roles in batteries, supercapacitors, fuel cells, redox flow batteries, electrolysis, and water splitting. The interplay between membrane structure, ion transport, and device performance is discussed, along with emerging concepts such as stimuli-responsive and biomimetic membranes. By combining fundamental science with technological perspectives, this book aims to serve as a useful resource for researchers and engineers working towards the development of high-performance electrochemical energy systems.

**Dr. G. Arthanareeswaran Ph.D.,**
National Institute of Technology, Tiruchirappalli

**Dr. Pei Sean Goh Ph.D.,**
Universiti Teknologi Malaysia

**Mr. Hemanth Kumar K MTech**
National Institute of Technology, Tiruchirappalli

# Funding Acknowledgement

The authors, G. Arthanareeswaran and Hemanth Kumar K, gratefully acknowledge the financial support provided by the Scheme for Promotion of Academic and Research Collaboration (SPARC), Ministry of Education, Government of India, under SPARC Project No. P4036, titled "Sustainable Electrodialysis-Driven Separation and Resource Recovery in Biorefinery."

# Abbreviations

| | |
|---|---|
| **2D** | Two-dimensional |
| **AEMs** | Anion-exchange membranes |
| **AEMWE** | Anion exchange membrane water electrolysis |
| **AFCs** | Alkaline fuel cells |
| **AFM** | Atomic force microscopy |
| **AIEMs** | Aromatic ion-exchange membranes |
| **AWE** | Alkaline water electrolysis |
| **CEMs** | Cation-exchange membranes |
| **CNT** | Carbon nanotubes |
| **CO** | Carbon monoxide |
| **COFs** | Covalent organic frameworks |
| **CTFs** | Covalent triazine frameworks |
| **DLP** | Digital light processing |
| **DMFCs** | Direct methanol fuel cells |
| **EDLCs** | Electric double-layer capacitors |
| **EIS** | Electrochemical impedance spectroscopy |
| **ESR** | Equivalent series resistance |
| **ESW** | Electrochemical stability window |
| **FDM** | Fused deposition modelling |
| **GPEs** | Gel polymer electrolytes |
| **HFP** | Hexafluoropropylene |
| **HPAs** | Heteropolyacids |
| **HT-PEMFCs** | High-temperature proton-exchange membrane fuel cells |
| **HT-PEMs** | High-temperature proton-exchange membranes |
| **IEC** | Ion exchange capacity |
| **LbL** | Layer-by-Layer |
| **LDHs** | Layered double hydroxides |
| **LIBs** | Lithium-ion batteries |
| **MCFCs** | Molten carbonate fuel cells |
| **MEA** | Membrane–electrode assembly |
| **MMMs** | Mixed-matrix membranes |
| **MOFs** | Metal-organic frameworks |
| **·OH** | Hydroxyl |
| **·OOH** | Peroxide |
| **ORR** | Oxygen reduction reaction |
| **PA** | Phosphoric acid |
| **PAEK** | Poly(aryl ether ketone) |
| **PAFCs** | Phosphoric acid fuel cells |
| **PANI** | Polyaniline |
| **PBI** | Polybenzimidazole |
| **PE** | Polyethylene |
| **PEM** | Proton exchange membrane |

| | |
|---|---|
| **PEMELs** | Proton exchange membrane electrolyzers |
| **PEMFCs** | Polymer electrolyte membrane fuel cells |
| **PEMWE** | Proton exchange membrane water electrolyzers |
| **PEMWE** | Proton exchange membrane water electrolysis |
| **PEO** | Polyethylene oxide |
| **PFSA** | Perfluorosulphonic acid |
| **PI-g-PEO** | Polyimide-grafted polyethylene oxide |
| **PP** | Polypropylene |
| **PPO** | Polyphenylene oxide |
| **PPy** | Polypyrrole |
| **PTFE** | Polytetrafluoroethylene |
| **PVDF** | (Poly(vinylidene fluoride) |
| **RFBs** | Redox flow batteries |
| **RH** | Relative humidity |
| **ROS** | Reactive oxygen species |
| **SEM** | Scanning electron microscopy |
| **SIB** | Sodium-ion batteries |
| **SLA** | Stereolithography |
| **$-SO_3^-$** | Sulphonate |
| **SOFCs** | Solid oxide fuel cells |
| **SPEEK** | Sulphonated polyether ether ketone |
| **SPEs** | Solid polymer electrolytes |
| **SPI** | Sulphonated polyimides |
| **SPPSU** | Sulphonated poly(phenylene sulfone) |
| **SPSU** | Sulphonated polysulfone |
| **SSBs** | Solid-state batteries |
| **TEM** | Transmission electron microscopy |
| **TGA** | Thermogravimetric analysis |
| **TIPS** | Thermally induced phase separation |
| **VIPS** | Vapour-induced phase separation |
| **VRFBs** | Vanadium redox flow batteries |
| **VRFBS** | Vanadium redox-flow batteries |

# 1 Introduction to Membranes in Electrochemical Devices

## 1.1 INTRODUCTION

Electrochemical devices play a crucial role in future energy systems for efficient energy storage and conversion, which is an imperative for sustainable development [1]. This class of devices, including fuel cells (FCs) and batteries, involves the discharge of electrons through one or more redox (reduction–oxidation) processes at the electrode–electrolyte interface and simultaneously reverses the energy-storage process, as seen in electrolyzers and rechargeable batteries [2]. Their efficiency, reliability, and scalability have triggered a global shift in the world towards low-carbon energy infrastructure and have been widely adopted. The efficiency of these devices relies on the membrane, which is a selective porous material that regulates the transport of ions, insulates electrons, and ensures reactive species are separated physically. Through the selective permeability of certain ions, like hydrogen ions ($H^+$), hydroxide ions ($OH^-$), lithium ($Li^+$), or vanadium ions, while blocking gases, electrons, or other undesired substances from entering the membranes, the system's efficiency, durability, and safety can be enhanced [3]. In addition to ionic transport, membranes also help in regulating water balance, heat response, and mechanical strength, making them multifunctional materials that integrate elements of electrochemistry, mechanics, and materials science.

The defining characteristics of electrochemical devices vary with the specific application, but almost all include the membrane–electrode assembly (MEA) concept, where the membrane is in direct contact with the electrodes and the electrolyte. Examples are given below:

1. **Fuel cells**: The proton-exchange or anion-exchange membranes allow the transport of ions in the solid material. They are processed to fuel, and the oxidizer is separated.
2. **Batteries**: The microporous membranes or solid electrolytes control ion pathways, suppress dendrite formation, and provide thermal barrier safety.
3. **Supercapacitors**: The thin and porous membranes provide assurance of the absence of short circuits between electrodes while facilitating the fast passage of ions through the electrolyte. The dependence of supercapacitors on membrane porosity, wettability, and ionic accessibility is very high.

DOI: 10.1201/9781003679363-1

4. **Redox flow batteries**: The ion-exchange membranes ensure the exclusion of active species while maintaining charge neutrality in liquid electrolytes.
5. **Electrolyzers**: The membranes permit ion transport while tolerating harsh oxidative and alkaline environments, as found in hydrogen production.

Selecting a suitable membrane for each device depends on specific requirements such as ionic conductivity, selectivity, chemical stability, mechanical strength, and cost. Gaining a fundamental understanding of these aspects lays the groundwork for exploring device-specific membrane technologies in subsequent chapters.

### 1.1.1 Fuel Cells

FCs are devices that use an external fuel as a source of chemical energy to produce electrical energy via electrochemical reactions with an oxidant. Compared to combustion-based systems, FCs provide high power density, high efficiency, practically zero waste, and small portable equipment for large-scale power generation [4]. The most important part of the FC is the MEA, which consists of two parts, namely the membrane and the electrocatalyst. The electrochemical reaction will be done in this part. The role of the membrane between the electrodes is to conduct the protons produced at the anode to the cathode, inhibiting crossover of fuel and oxidant. FCs are classified based on operating conditions and FC structure. Therefore, based on this classification, FCs include proton-exchange membrane Fuel Cells (PEMFC), alkaline fuel cells (AFC), phosphoric acid fuel cells (PAFC), molten carbonate fuel cells (MCFC), and solid oxide fuel cells (SOFC).

PEMFCs are electrochemical devices that operate at 60–120°C, utilizing perfluorosulfonic acid (PFSA) membranes, such as Nafion, for proton conduction. They provide high power density up to 350mW/cm², fuel efficiency of about 45–60%, and are suitable for portable, mobile, and stationary applications. However, PEMFCs face some challenges, including the need for external humidification, loss of some mechanical and proton-moving qualities over 90°C due to membrane degradation, and cost challenges in the membrane and catalyst materials [5]. AFCs employ alkaline electrolytes like potassium hydroxide and work via hydroxide ion conduction and facilitate the use of non-precious metal catalysts; AFCs normally function at temperatures of 60–90%, power densities between 100 and 200 mW/cm², and have 40 to 60% efficiency. SOFCs are ceramic inorganic oxide as the electrolyte, usually yttria-stabilized zirconia rather than a liquid electrolyte, and require operation at elevated temperatures, typically between 750 and 1,000°C, and the power densities are around 240 mW/cm². They generally run on a mixture of hydrogen and carbon monoxide formed by internally reforming a practical hydrocarbon fuel within the FC, and use air as the oxidant, with the production of water and carbon dioxide [6]. The main challenges with these technologies are material degradation from thermal cycling and the high cost of ceramic components. PAFCs have a phosphoric acid electrolyte and run on hydrogen fuel with air or pure oxygen as the oxidant, and have an operating temperature of up to 200°C. Unlike the AFC, PAFCs are capable of operating on hydrogen contaminated with $CO_2$, which places less stringent demands on the external fuel processor [7]. MCFC have the molten potassium lithium carbonate electrolyte, which requires an operating temperature of around 650°C, with power densities around 100 $mW/cm^2$ and fuel efficiencies of 60–65%. They run on a mixture of carbon monoxide and hydrogen, that is, synthesis gas, which is formed by internally reforming practical hydrocarbon fuels within the FC. These systems can achieve lifetimes beyond 40,000 hours, with capital costs of approximately \$1000/kW, and are mainly used for distributed power generation. The role of membranes is to define the ion-transport mechanism (proton, hydroxide, oxide, carbonate), establish the operational temperature window, and influence the type, durability, and scalability of FCs. Better membranes are going to be extremely important in moving FCs from something used only in limited cases to a generally used, clean energy technology.

### 1.1.2 Batteries

Batteries are device that converts chemical energy contained within their active materials directly into electric energy by means of an electrochemical oxidation-reduction (redox) reaction. This type of reaction is carried out by the transfer of electrons from one material to another via an electric circuit. (Figure 1.1). During discharge,

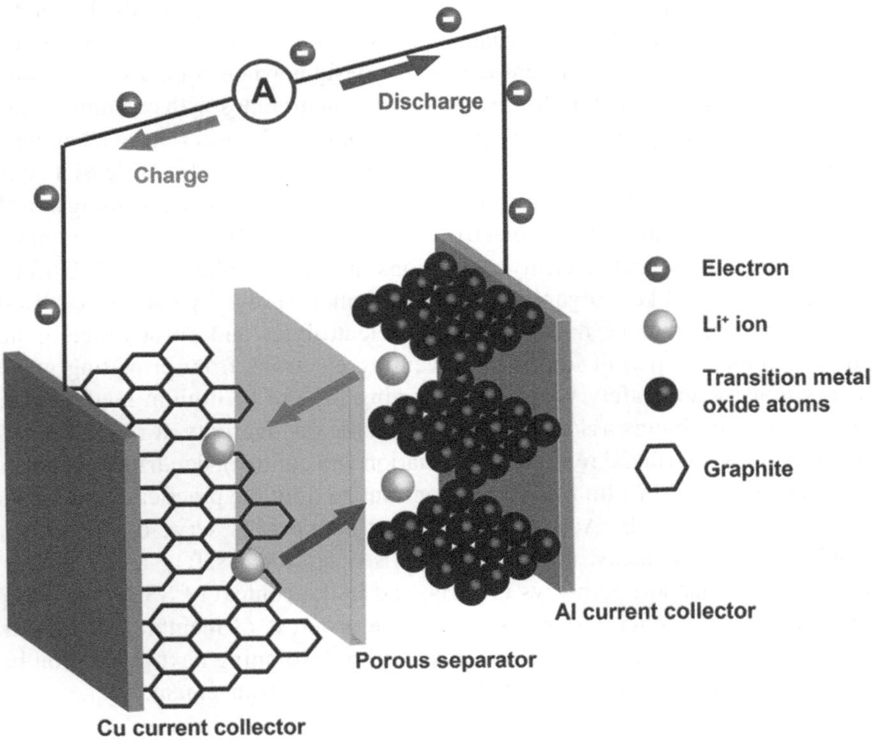


**FIGURE 1.1** Schematic representation of major components in LIBs.

(Reproduced with permission [8].)

electrons pass through the external circuit while ions move through the membrane–electrolyte system to balance the charge. During recharging, the operations reverse direction, allowing batteries to be reused multiple times. The design and selection of the membrane are crucial because they govern ionic transport, prevent electron passage, and influence safety, efficiency, and the device's lifespan [9].

Rechargeable batteries are considered the main energy sources due to their high specific capacities, long cycle lives, and sustainable properties, and lithium-ion batteries (LIBs) are ideal energy-storage systems due to their high energy densities, long cycle lives, and excellent chemical and mechanical properties. In conventional LIBs, microporous polymer membranes made of polyethylene (PE), polypropylene (PP), or multilayer polyolefin structures are the main membrane components. Their purpose is to isolate the anode and cathode to avoid electrical shorts and to enable lithium-ion migration through the liquid electrolyte absorbed in their pores [10]. Major membrane characteristics consist of porosity (to supply sufficient ionic conductivity), mechanical strength (to resist electrode pressure), and thermal shutdown behaviour (to improve safety under overheated conditions) [11]. These coatings include ferric oxide, manganese oxide, and phosphoric acid (PA), as well as ceria-cobaltate, alumina ($Al_2O_3$) for lithium-ion battery membranes, or Silicon dioxide ($SiO_2$)-embedded membrane. They also offer thermal stability against meltdown when overheated and enhanced dendrite resistance to prevent short circuiting from metal growth. However, because of the larger $Na^+$ radius and different solvation dynamics of $Na^+$ ions in the electrolyte, sodium-ion batteries (SIB) need different membrane designs when compared to lithium-ion systems. Even though traditional polyolefin membranes with customized pore structure can satisfy the basic requirements, they still suffer from challenges, including reduced cycle life and electrolyte incompatibility. Feasible SIB centre around interfacial stability between cathode and anode materials alongside the use of a suitable sodium-based electrolyte. A solid-state battery (SSB) combines both the membrane and electrolyte functions in a single solid material. Different solid electrolytes, like inorganic ceramics, polymers (polyethylene oxide (PEO)), and composites, replace flammable liquid electrolytes and porous membranes. These membranes permit ion flow, preventing electrons from providing greater energy density and safety while also allowing the use of lithium metal anodes. Nonetheless, problems associated with low ionic conductivity at room temperature as well as interfacial resistance (formation and control), dendrite penetration, and scalability in thin-film fabrication continue to limit the practical implementation of these materials. Accordingly, battery membranes serve double-duty as enablers and bottlenecks: they enhance LIB operational safety and efficiency; redesign ion-transport pathways in SIBs; and in the context of SSBs, they shape the future of safe, energy-dense storage. The progress of membrane chemistry, morphology, and processing will be essential in developing next-generation battery technologies and determining their commercial advancement.

### 1.1.3 Supercapacitors

Supercapacitors are generally described as energy-storage devices that lie between conventional capacitors and batteries. They are mainly known for their ability to charge and discharge quickly while still storing a reasonable amount of energy. Because of this, they have received considerable research attention. Compared to batteries, supercapacitors usually show longer cycle life, higher power density, and better operational safety. The energy-storage process is also highly efficient since it mainly involves reversible ion adsorption at the electrode surface rather than slow bulk chemical reactions. The energy-storage mechanism in supercapacitors differs from that in batteries. In electric double-layer capacitors, energy is stored through electrostatic charge separation at the interface between the electrode and electrolyte as shown in Figure 1.2. In pseudocapacitors, fast surface redox reactions also contribute to charge storage. Since these processes occur near the surface, the devices can deliver rapid energy output. The membrane or separator between the electrodes is an important component because it controls ion movement within the device.

The membrane primarily prevents direct physical contact between the electrodes while still allowing ions to pass through the electrolyte. In many commercial systems, microporous polymer membranes made from PE or PP are commonly used. These materials provide mechanical strength and chemical resistance, and they can retain the liquid electrolyte inside their pores. However, one limitation is that their hydrophobic nature sometimes reduces electrolyte wettability, particularly when ionic

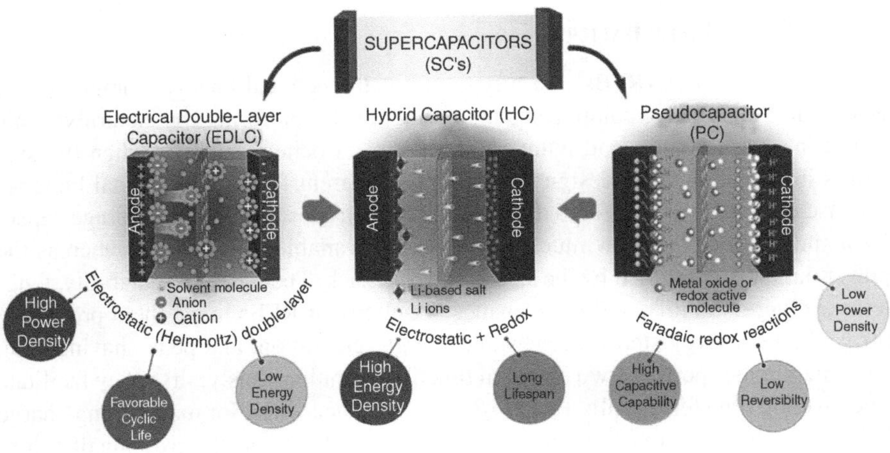


**FIGURE 1.2** Schematic representation of conventional supercapacitors (SCs). **(Reproduce with permission [12].)**

liquids or gel electrolytes are used, which may influence the electrochemical behaviour. To improve performance, researchers have explored gel polymer electrolytes and solid polymer electrolytes as alternative membrane systems. Polymers such as Poly(vinylidene fluoride) (PVDF), Poly(vinylidene fluoride-hexafluoropropylene) (PVDF-HFP), and PEO are often selected because they can provide flexibility along with electrochemical stability [13]. When these polymers are combined with salts or ionic liquids, they can support ion conduction while also acting as structural separators. Such materials may help in lowering internal resistance and can also contribute to safer device configurations, especially in flexible or solid-state designs.

Recently, there has also been interest in environmentally friendly membrane materials. Natural polymers, paper-based structures, and bacterial cellulose membranes have demonstrated high porosity and strong electrolyte-absorption capacity. In addition, nanocomposite membranes prepared by electrospinning or by adding nanomaterials such as graphene oxide or metal oxides have demonstrated improved ionic transport and cycling behaviour. These modifications often create more efficient ion pathways and improve interactions between components within the device [14]. Overall, the separator membrane plays a significant role in determining the electrochemical performance of supercapacitors. It influences ion transport, internal resistance, mechanical stability, and long-term durability. Parameters such as capacitance, rate capability, leakage current, and cycle stability are therefore closely related to membrane properties. Because supercapacitors depend on fast ion movement, membranes used in these systems must maintain good ionic conductivity together with sufficient mechanical and thermal stability for reliable operation.

### 1.1.4 Redox Flow Batteries

Redox flow batteries (RFBs) are a type of electrochemical energy-storage system where the electroactive components are in liquid solution form. The electrolytes are stored in external tanks and pumped into the electrochemical cells when the system is in operation. This design makes RFBs different from conventional batteries because the energy is not stored directly in the electrodes. Instead, the storage capacity mainly depends on how much electrolyte is available in the tanks, whereas the power output is controlled by the size of the cell stack. The inherent scalability, long-term cycling durability, and maintenance simplicity of RFBs make them promising for grid-level energy storage, renewable energy integration, and peak shaving [15]. RFB membranes perform two essential functions simultaneously: first, they facilitate selective ion transfer (usually $H^+$, $SO_4^{2-}$, or $OH^-$) necessary for maintaining charge balance during redox reactions; second, they physically restrict the crossing of redox-active species (e.g., $V^{2+}/VO^{2+}$ in vanadium RFBs, $Fe^{2+}/Fe^{3+}$ in iron-based systems). Excessive crossover of electroactive species results in self-discharge, capacity fading, and reduced coulombic efficiency, which makes ion selectivity and chemical stability key considerations in membrane design. The rational selection of a suitable material is essential to suppress crossover while maintaining high proton conduction and water swelling tolerance. The systematic optimization of layered solid-acid materials has emerged as a promising strategy for developing next-generation membranes. These materials exhibit the potential to deliver enhanced proton conductivity, excellent water swelling tolerance, and superior selectivity, making them strong candidates for future RFB membranes. Currently, PFSA membranes, such as Nafion, set performance standards due to their high proton conductivity and long-term stability [16]. Despite their advantages, their high cost and relatively poor resistance to vanadium ion crossover have driven research efforts towards alternatives such as hydrocarbon-based polymers (e.g., sulfonated polyether ether ketone, SPEEK), anion-exchange membranes (AEMs), and microporous membranes with specifically tailored pore sizes.

Innovative strategies integrating composite technology with nanoparticles, graphene oxide, or inorganic fillers have garnered extensive research interest. These methods have always proved their potential to lower species permeability with retaining ionic conductivity. In order to resist the corrosive conditions in an RFB, membranes are required to possess certain physicochemical properties, among which chemical resistance to oxidation is a major concern. In addition, long-term durability is essential, requiring membranes to withstand continuous mechanical stresses caused by electrolyte flow and pressure fluctuations. An effective membrane must achieve an optimal balance among high ionic conductivity, selectivity, mechanical strength, coulombic efficiency, and to ensure durability during continuous operational conditions. Membranes determine the efficiency, longevity, and economic viability of RFBs.

### 1.1.5 Electrolyzers

Electrolyzers are electrochemical devices that split the water into hydrogen and oxygen using electricity, and in modern devices, this is achieved through the membrane or the electrolyte that is able to conduct ions, separate the gases, and provide electrical insulation between the electrodes and the anode and cathode. Hence, completing the ionic circuit provides the path through which the electrons must pass through the external circuit. In low-temperature electrolyzers, polymer membranes dominate, especially the proton-exchange membrane in PEM electrolyzers and the anion-exchange membrane in alkaline polymer electrolyzers, compared to the conventional liquid KOH with porous diaphragms and the dense ceramic electrolyte in the case of the solid oxide electrolyzers [17].

The fundamental functions of these membranes in electrochemical systems are: (i) to offer high ionic conductivity for the target ions, namely, $H^+$ in PEM, $OH^-$ in AEM or Alkaline Water Electrolysis (AWE), and $O^{2-}$/vacancies in solid oxide electrolysis cells (SOEC); (ii) to prevent mixing of hydrogen and oxygen gases by functioning as a physical separator, thereby ensuring high purity of the gases and avoiding the possibility of explosions; and (iii) to prevent electrical conduction, thereby avoiding short circuits.

To achieve the necessary functions, the membrane must be permselective, allowing high ionic conductivity for target ions while blocking other ions and electrons. It needs to have low gas crossover to maintain high Faradaic efficiency and must also be mechanically durable to endure compression, differential pressure, and hydration/

dehydration cycles. It should also be chemically and thermally stable within the device environment, resisting oxidation and hydrolysis in strongly acidic PEM systems, or degradation and carbonation in AEM systems. Additionally, it must maintain high ionic conductivity, usually around 0.1 S/cm for advanced PEM systems. Different water electrolysis technologies are categorized based on their membrane or electrolyte types. For instance, alkaline water electrolyzers utilize liquid KOH as the electrolyte and a porous diaphragm to conduct $OH^-$ ions. This approach offers reliable, mature hardware at a lower cost, though it typically operates at lower current densities. PEM electrolyzers use a solid PFSA-type membrane such as Nafion for $H^+$ conduction, which provides high-current density and high pressure with very high hydrogen purity but requires noble metal catalysts and uses relatively expensive fluorinated polymers. AEM electrolyzers substitute the acidic PEM with an anion-conducting polymer with functionalized cations (e.g., quaternary ammonium), enabling $OH^-$ conduction in an alkaline medium and paving the way for non-precious metal catalysts, though still plagued by durability and long-term conductivity SOECs utilize dense oxide ceramics (e.g., YSZ) for the conduction of oxide ions or oxygen vacancies at high temperatures, offering high thermodynamic efficiency but more complex material requirements.

In PEM technology, PFSA membranes are now the industry standard because of their phase-separated microstructure, with hydrophilic, sulfonic acid-rich nanochannels embedded in a fluoropolymer matrix, enabling high proton conductivity and durability. In alkaline technology, polymers functionalized with quaternary ammonium groups, embedded within a hydrophobic fluorocarbon backbone, enhance high proton mobility and exhibit strong mechanical properties. New hydrocarbon-based PEMs and advanced AEMs are in development to reduce material costs, expand operational condition ranges, and improve compatibility with various catalysts.

## 1.2 ROLES OF MEMBRANE AND SEPARATION MECHANISMS

In electrochemical devices, membranes act as essential components that directly influence electrochemical efficiency, device longevity, and operational safety. Their overall performance results from a complex interplay of factors, including physical structure, chemical composition, and ion-transport properties. As illustrated in Figure 1.3, ion-transport behaviour during charging and discharging in Li-ion systems highlights the critical role of membranes in governing electrochemical performance. **Ion conductivity**: the membrane facilitates the selective transport of specific charge carriers depending on the electrochemical system, such as $H^+$ in proton-exchange membranes (PEMs) FCs systems, $OH^-$ in AEM systems, $Li^+$ or $Na^+$ in batteries, and supporting ions such as $SO_4^{2-}$ in flow batteries. This establishes an almost one-to-one correspondence between hydrogen ions and electrons through redox reactions, enabling the formation of an ionometric hybrid system that outperforms conventional electrolytes. The membrane must suppress the crossover of undesired species to prevent capacity fade, low voltage efficiency, and reduced lifetime. **Selective barrier**: the membrane selectively allows ions to pass through while blocking impurities and larger molecules. For example, PEMs are used to

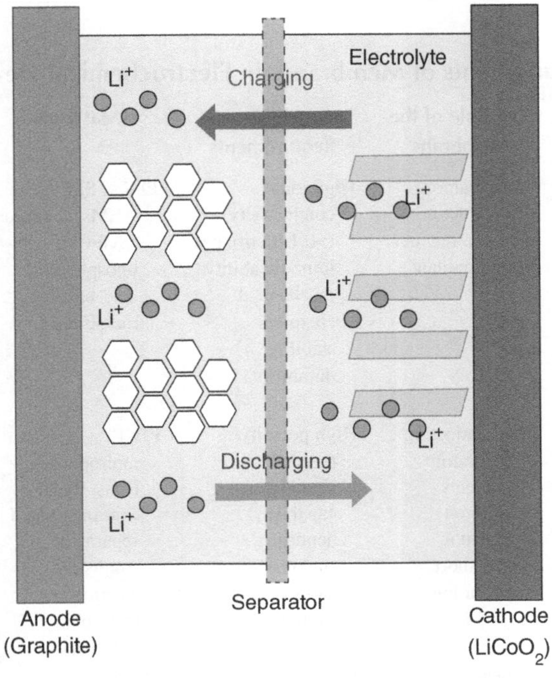


**FIGURE 1.3** Schematic illustration of the structure and ion-conducting mechanism of Li-ion batteries.

**(Reproduced with permission [18].)**

transport protons from the anode to the cathode by preventing the crossover of fuel and oxidant. In batteries, it prevents electrode-damaging species from leakage and solvent mixing. This high selectivity improves efficiency and reduces issues like short circuits and self-discharge. **Charge balance**: electrons are allowed to reach the other electrode only through the external wire that connects the two half-cells of the electrochemical cell; meanwhile, ions with opposite charges pass through the electrolyte across the membrane, maintaining electroneutrality in both electrodes. Maintaining separate paths of electrons and ions ensures efficient, sustained, and reversible operation. **Mechanical strength**: the membrane physically separates the electrodes and must endure pressure, swelling caused by the electrolyte, heat, and repeated cycling. It needs to maintain its structural integrity and resist deterioration due to these factors. Collectively, these functions emphasize the membrane's crucial role in regulating ion transport and maintaining system stability. Instead of acting as passive barriers, membranes function as multifunctional interfaces that integrate ionic conduction, chemical selectivity, and structural robustness, thereby enhancing the efficiency, durability, and safety of electrochemical devices. The key roles and requirements of membranes in electrochemical devices are summarized in Table 1.1.

**TABLE 1.1**
**Role and Requirements of Membranes in Electrochemical Devices**

| Electrochemical Device | Key Role of the Membrane | Key Requirements | Materials | Major Challenges |
|---|---|---|---|---|
| **Fuel Cells** | Proton/anion conductor; separates fuel and oxidant | High ionic conductivity (>0.1 S/cm), gas impermeability, mechanical/chemical stability, durability | PFSA, SPEEK, AEMs, ceramics (SOFCs), phosphoric acid-based membranes | High cost of PFSA, durability at high T, $CO_2$ sensitivity (AFCs), thermal cycling (SOFCs) |
| **Batteries (LIBs, SIBs, SSBs)** | Ion-conducting separator prevents electron transport and short circuiting | High porosity, thermal shutdown capability, dendrite suppression, and mechanical strength | PE, PP microporous films, PVDF, ceramic-coated separators, solid-state electrolytes (ceramic, polymer) | Thermal safety, dendrite growth, low room-T conductivity (SSBs), interfacial resistance |
| **Supercapacitors** | Prevents electrode shorting while allowing ion mobility | High porosity, electrolyte wettability, low ESR, and mechanical flexibility | PP, PE membranes, PVDF, PEO-based GPEs, bacterial cellulose, graphene oxide composites | Poor wettability of hydrophobic membranes, leakage current, trade-off between strength and porosity |
| **RFBs** | Selective ion transport prevents crossover of redox species | High proton/ionic conductivity, low permeability to redox ions, chemical stability, and mechanical durability | PFSA, SPEEK, AEMs, microporous composites, inorganic–polymer hybrids | High cost, vanadium crossover, swelling, oxidative degradation |
| **Electrolysers** | Conduct ions between electrodes, separate $H_2$ and $CO_2$ gas to ensure high purity | High ionic conductivity, low permeability, and chemical stability | PEM: Nafion, SPEEK<br>AEM: PSU, PBI | High cost, gas crossover at low current densities, membrane degradation under pressure |

## 1.3 CHRONOLOGICAL DEVELOPMENT OF MEMBRANES FOR ELECTROCHEMICAL DEVICES

The development of membrane technology over several decades is linked to advances in materials science, electrochemistry, and market demand.

**1890–1920s**: Membrane science began with early discoveries. In 1890, Wilhelm Ostwald observed the membrane potential. In 1911, Frederick Donnan described what is now called the Donnan equilibrium. By 1925, homogeneous weakly acidic membranes were made using wool adhesives, marking the start of practical membrane development.

**1930s–1940s**: Membrane technology advanced with the introduction of charged membranes. Phenol-formaldehyde ion-exchange membranes were developed. Electrodialysis was also introduced, enabling early selective ion-transport applications.

**1950s–1970s**: Commercial ion-exchange membranes were first produced by Juda. Winger developed low-resistance, selective membranes that made large-scale electrodialysis feasible. In the 1960s, PEMFC were developed for space applications, and PEM water electrolysis started. The 1970s saw the commercialization of Nafion membranes, which offered phase-separated ionic channels for high proton conductivity and chemical stability.

**2000s–2010s**: PEMFC achieved operational lifetimes of about 2,500 hours. Research focused on water and thermal management and cost reduction. At the same time, AEMs became more widely studied because they worked with non-precious metal catalysts and showed improved stability and power density.

**2020s–future**: New ion-exchange membranes with engineered microporous channels were developed. Reinforced poly(arylene alkylene) AEMs showed durability beyond 1,000 hours. Hybrid PEM-AEM membrane–electrode assemblies were introduced to lower costs and enhance performance, demonstrating continued membrane evolution for modern electrochemical devices.

## 1.4 KEY MEMBRANE PARAMETERS

Membranes are essential separators in electrochemical devices, and their functions depend on several interconnected parameters such as ionic conductivity, mechanical stability, and chemical resistance.

### 1.4.1 Ionic Conductivity (S/cm)

Ionic conductivity contributes to ion transport across the membrane and, in turn, affects the internal resistance of a cell. The conductivities of the PEM membranes are uniformly higher, and it is anticipated to be in the range of 0.1 S/cm, which promotes high-current densities for operation. For AEM systems and some solid electrolytes, the conductivity values are typically much lower, though they can be enhanced through propagation of thinner membranes or improved electrode–membrane contact. The hydration of the membrane, temperature, and structure of the polymer are important factors for conductivity. In PEM membranes, protons travel through hydrated pathways, while in AEM membranes, hydroxide ions move along the functional cationic groups [19]. Recent studies have also shown that various head groups, including quaternary ammonium and similar structures, can enhance conductivity and alkaline stability for AEM-based FCs and electrolyzers.

### 1.4.2 Selectivity and Permeability

Membrane selectivity refers to the ability of the membrane to allow the required ions to pass while restricting other species [20]. High permeability of undesired species results in severe crossover, leading to rapid performance degradation and reduced coulombic efficiency. In PEMFC, this is mainly related to reducing hydrogen and oxygen crossover, since gas mixing can lower efficiency. In redox flow battery systems, ion crossover through the membrane is associated with self-discharge and capacity loss during cycling. If the permeability of unwanted species is high, performance degradation may occur more quickly. Reported approaches to improve selectivity include structural modification of the polymer, crosslinking, and preparation of reinforced or composite membranes to create more difficult transport pathways for undesired species [21].

### 1.4.3 Mechanical Strength

Ion-exchange membranes, like PEMs and AEMs, are subjected to various mechanical stresses during their operation. The sources of these stresses are primarily due to the compression of the stack, hydration and dehydration cycles, and temperature variation that lead to expansion or contraction. When the mechanical strength is low, cracks may develop, which can lead to delamination, gas leakage, or internal short circuits. In AEM, the membrane is exposed to hydroxide ions alkaline environment, where they may damage the ionic group attached to the polymer or the polymer backbone. PEMs may also experience stress, especially when they are used under pressurized conditions. To enhance the strength of the membrane, support materials like porous PTFE or fabric layers may be incorporated, and crosslinking of the polymer chain is also possible. These methods assist in maintaining the stability of the membrane.

### 1.4.4 Chemical and Electrochemical Stability

The membranes are exposed to severe chemical environments and electrochemical reactions. The PEMs need to maintain their structure in an acidic environment and be stable against radicals generated during the oxygen evolution reaction. The stability is also critical for AEMs, which are susceptible to nucleophilic attack, polymer degradation through functional group substitution, and carbonate formation under alkaline conditions. Redox flow battery membranes face highly oxidizing and reducing species sequentially. It is essential that all functional groups, polymer chains, and fillers be highly stable. For instance, a stress test, such as Fenton's reagent for PEMs, which is used to accelerate degradation, is a common method to evaluate chemical stability. Similarly, high-pH soak tests are used to test the chemical stability of AEMs.

### 1.4.5 Water Uptake and Swelling

Water uptake affects the ion-transport behaviour of the membrane. In PEM membranes, hydration is necessary for proton conduction, and in AEM membranes, water is involved in hydroxide ion transport. At the same time, higher water absorption can lead to membrane swelling and dimensional changes. This may cause mechanical stress at the membrane–electrode interface. Reported methods to control water uptake include polymer modification, crosslinking, and the use of composite materials. Swelling of the membrane can also influence gas permeability and device performance.

### 1.4.6 Thermal Stability

Apart from ionic conductivity, the membrane should also remain stable within the operating temperature range of the device. PEMFC typically operate at around 80°C, whereas AEM and solid electrolyte systems may be exposed to higher temperatures. Thermal stability during cold start depends on the competition between heat generation and ice formation. Parameters related to the MEA, including membrane thickness, influence temperature distribution and current density during startup. These transport and reaction interactions determine whether the FC can successfully start under subzero conditions.

### 1.4.7 Efficiency

The efficiency of electrochemical devices is closely related to the membrane properties, particularly ionic conductivity and thickness. In PEM systems, membranes with higher proton conductivity generally show lower internal resistance and better cell performance. Nafion membranes are commonly used due to their high conductivity (0.1 S/cm), although alternative materials such as SPEEK have been used, they require high water content to match the conductivity of Nafion. In AEM systems, studies indicate that membrane thickness and water management influence the achievable current density and power density. Operating at high temperature can reduce membrane resistance, but optimized management can prevent swelling and maintain mechanical strength.

### 1.4.8 Energy Density

Membrane thickness and conductivity affect the device's energy density. A thinner membrane decreases the distance for ion-transport pathways and lowers resistance (ohmic losses). This can improve the ionic conductivity, voltage efficiency, and better utilization of active materials. These factors become more significant in portable systems and high-energy-density batteries where space and weight are limited.

### 1.4.9 Durability and Safety

Membrane stability affects both the lifetime and safety of electrochemical devices. Chemical resistance and mechanical strength are needed to maintain performance during operation. Membranes also act as barriers that reduce gas crossover and internal short circuits. Degradation of the membrane may lead to performance loss or safety concerns. For this reason, optimization of membrane materials and structure is widely studied for reliable device operation [22].

## 1.5 CHARACTERIZATION TECHNIQUES

The various methods are used for the thorough characterization of membrane performance. Ion-exchange membrane conductivity and resistance can be directly measured using electrochemical impedance spectroscopy (EIS), while thermogravimetric analysis (TGA) provides information about a membrane's thermal degradation and stability. Scanning electron microscopy (SEM) and Atomic force microscopy (AFM) imaging are very helpful in revealing specific surface characteristics of materials, such as pore size and phase separation. The density of functional groups on the membrane surface is determined through ion-exchange capacity (IEC) measurements, with the IEC value exhibiting a direct correlation to the membrane's ionic conductivity [23]. Additionally, mechanical tests like static tensile tests, puncture tests, and fatigue cyclic loading offer valuable insights into the polymer's mechanical properties; the results can be used to assess the polymer's stability. Gas permeability tests are important for evaluating membrane gas crossover behaviour, which is critical for safety and efficiency. The main advantage of membrane technologies is their ability to combine essential properties such as high ionic conductivity, selectivity, mechanical robustness and stability, water management, and thermal tolerance into a single device. Achieving high-performing, durable, and safe electrochemical devices depends on the synergistic optimization of these factors, often through advanced polymer chemistry, composite design, and structural engineering.

### 1.5.1 Electrochemical Impedance Spectroscopy

EIS is a non-destructive technique used to analyse the electrical properties of an electrochemical system across a range of frequencies. It is based on electrical impedance measurements performed using an alternating sinusoidal current over a frequency range of $10^{-6}$~$10^{9}$ Hz. EIS has been successfully applied to study electrochemical processes at different time scales, including the characterization of multilayer systems, material dielectric and transport properties, and electrochemical reaction mechanisms. Since EIS renders the individual assessment of the electrical contribution of each sublayer available, it has been applied to characterize synthetic IEMs, composite membranes, and bipolar membrane structures.

### 1.5.2 Thermogravimetric Analysis

TGA is used to evaluate thermal stability by monitoring mass change as a function of temperature in ion-exchange membranes, a critical property for their reliable operation in electrochemical environments that involve elevated temperatures. It identifies key thermal transition points, water loss, functional group degradation, and backbone decomposition critical for ensuring membrane reliability in high-temperature electrochemical applications, such as FCs and water electrolyzers.

### 1.5.3 Scanning Electron Microscopy (SEM)

SEM is used to observe the detailed morphological structure of ion-exchange membranes, providing necessary structural information that depends on ion-transport behaviour and membrane performance. This characterization is used to identify the different phases or particular domains, such as ion-exchange regions and inert binder, and to assess porosity, pore size distribution, and irregular structural features of the membrane, which can influence the ionic conductivity and mechanical properties of the membrane.

### 1.5.4 Atomic Force Microscopy

AFM is widely used to characterize ion-exchange membranes due to its high-resolution imaging, 3D surface topography, and quantitative surface roughness analysis, performed in the absence of vacuum conditions or extensive sample preparation, enabling detailed evaluation of membrane surface morphology, phase separation, and internal nanostructural features that influence ionic transport and interfacial membrane performance. AFM can also map local variations in surface characteristics and interaction forces at the nanoscale to microscale, which are critical for understanding ion pathways, domain distribution, and the effects of membrane fabrication on functional performance.

## 1.6 CURRENT TRENDS IN MEMBRANE SCIENCE

The demand for higher performance, sustainability, and multifunctionality in electrochemical systems has spurred rapid advances in membrane research.

### 1.6.1 Membrane Materials and Modification

The use of functional fillers has emerged as a promising approach to enhance the conductivity, selectivity, and mechanical stability of PEMFC membranes. A polymer composite membrane operating at high temperatures can be improved by incorporating metal oxide, multi-walled carbon nanotubes, and reduced graphene oxide. Metal-organic frameworks (MOFs), particularly UiO-66-$NH_2$ and UiO-66-$SO_3H$, with sulfonated graphene oxide, have been identified as the most promising fillers for FC membranes. The composite membranes design not only enables a precise customization of membrane properties according to the requirements of the specific device but also enables an enhancement of performance along with durability. Solid-state ion conductors include polymer-ceramic hybrid membranes and inorganic solid electrolytes, such as garnet-type oxides and sulfide-based systems. These hybrid materials integrate the structural flexibility of polymers with the high ionic conductivity of ceramics, thereby fulfilling the essential criteria for thermally robust, thermally stable membranes suitable for high-current electrochemical applications. The use of solid-state conductors also aims to reduce leakage, improve safety, and enable high-energy-density devices.

The recent trend in the development of membrane based on improving their functionality. Examples of such membranes include self-healing polymers that can recover their mechanical properties after damage, temperature-responsive membranes whose water content depends on temperature, and redox-active membranes that, through electrochemical reactions, can slow down degradation. The inclusion of these multifunctional groups in membranes not only extends their operational lifespan without replacement but also enhances the reliability of devices under harsh working conditions [24].

### 1.6.2 Cost and Circular Economy

One of the main reasons for the high cost of PEMFCs is the use of fluorinated polymers like Nafion. This has driven the development of alternative materials based on hydrocarbons and bio-derived sources. To reduce the cost of high-performance membranes without affecting their qualities, polymer designs should be optimized and scalable production methods should be adopted. As a result, larger-scale applications in hydrogen production, batteries, and FCs will become more feasible. Membrane development is increasingly influenced by stability considerations. The factors for future membrane technology will include compatibility with sustainable energy goals, achieved through recyclability, minimal use of solvent, energy-efficient processing, and minimizing environmental impact. By adopting design-for-recycling principles, reducing toxic materials, and choosing materials that are easy to recycle or degrade at the end of their lifespan, researchers are progressing towards sustainable membranes.

## 1.7 CONCLUSION

In this chapter, the role of membranes in various electrochemical devices has been explained. Membranes play a major role not only in ion transport but also in separating reactants and maintaining the overall stability of the system. Their properties directly influence efficiency, durability, safety, and device lifetime. Depending on the type of device, such as FCs, batteries, supercapacitors, or RFBs, the membrane requirements may vary, and therefore, material selection becomes very important. The evolution of membrane materials has travelled from earlier materials like asbestos, cellulose, and glass fibre to more advanced polymer systems, like fluorinated membranes such as Nafion and hydrocarbon-based ionomers. Recently, there has also been growing interest in bio-based and eco-friendly membranes. In general, the membrane technology aims to improve conductivity, mechanical strength, chemical stability, and cost effectiveness for practical applications. Overall, membrane technology continues to improve along with the growing demand for efficient energy-storage and conversion systems. Understanding the relationship between membrane structure and performance is essential for designing better materials. The following chapters will discuss membrane classification, fabrication methods, and property–performance relationships in more detail, with particular focus on PEM, AEM, and solid-state membrane systems for advanced electrochemical applications.

## REFERENCES

[1] I. T. Bello, L. A. Jolaoso, R. A. Ahmed, and A. Bello, "Electrochemical energy conversion and Storage Systems: A perspective on the challenges and opportunities for sustainable energy in Africa," *Energy Rev.*, vol. 4, no. 1, p. 100109, 2025, doi: 10.1016/j.enrev.2024.100109

[2] S. Rashidi, N. Karimi, B. Sunden, K. C. Kim, A. G. Olabi, and O. Mahian, "Progress and challenges on the thermal management of electrochemical energy conversion and storage technologies: Fuel cells, electrolysers, and supercapacitors," *Prog. Energy Combust. Sci.*, vol. 88, no. October 2021, p. 100966, 2022, doi: 10.1016/j.pecs.2021.100966

[3] R. Pärnamäe et al., "Bipolar membranes: A review on principles, latest developments, and applications," *J. Memb. Sci.*, vol. 617, 2021, doi: 10.1016/j.memsci.2020.118538

[4] A. B. M. Ali, A. K. Nemah, Y. A. Al Bahadli, and Ehsan Kianfar, "Principles and performance and types, advantages and disadvantages of fuel cells: A review," *Case Stud. Chem. Environ. Eng.*, vol. 10, no. September, p. 100920, 2024, doi: 10.1016/j.cscee.2024.100920

[5] T. K. Maiti et al., "Advances in perfluorosulfonic acid-based proton exchange membranes for fuel cell applications: A review," *Chem. Eng. J. Adv.*, vol. 12, no. April, p. 100372, 2022, doi: 10.1016/j.ceja.2022.100372

[6] H. G. Desta, G. Gebreslassie, J. Zhang, B. Lin, Y. Zheng, and J. Zhang, "Enhancing performance of lower-temperature solid oxide fuel cell cathodes through surface engineering," *Prog. Mater. Sci.*, vol. 147, no. May 2024, p. 101353, 2025, doi: 10.1016/j.pmatsci.2024.101353

[7] K. Kargupta, S. Saha, D. Banerjee, M. Seal, and S. Ganguly, "Performance enhancement of phosphoric acid fuel cell using phosphosilicate gel based electrolyte," *Ranliao Huaxue Xuebao/Journal Fuel Chem. Technol.*, vol. 40, no. 6, pp. 707–713, 2012, doi: 10.1016/s1872-5813(12)60026-7

[8] T. Or, S. W. Gourley, K. Kaliyappan, A. Yu, and Z. Chen, "Recycling of mixed cathode lithium-ion batteries for electric vehicles: Current status and future outlook," *Carbon Energy*, vol. 2, no. 1, pp. 6–43, 2020, doi: 10.1002/cey2.29

[9] A. A. Sheikh, F. R. Bianchi, D. Bove, and B. Bosio, "A review on MCFC matrix: State-of-the-art, degradation mechanisms and technological improvements," *Heliyon*, vol. 10, no. 4, p. e25847, 2024, doi: 10.1016/j.heliyon.2024.e25847

[10] P. U. Nzereogu et al., "Solid-state lithium-ion battery electrolytes: Revolutionizing energy density and safety," *Hybrid Adv.*, vol. 8, no. November 2024, p. 100339, 2025, doi: 10.1016/j.hybadv.2024.100339

[11] M. Pigłowska et al., "Challenges for safe electrolytes applied in lithium-ion cells—a review," *Materials (Basel).*, vol. 14, no. 22, pp. 1–52, 2021, doi: 10.3390/ma14226783

[12] S. Q. Zheng, S. S. Lim, C. Y. Foo, C. Y. Haw, W. S. Chiu, C. H. Chia, and P. S. Khiew, "Recent progress on the applications of carbonaceous and metal-organic framework nanomaterials for supercapacitors," *Front. Mater.*, vol. 8, p. 777149, 2021, doi: 10.3389/fmats.2021.777149

[13] C. Han et al., "Recent progress in gel polymer electrolyte for lithium metal batteries," *Giant*, vol. 20, p. 100337, 2024, doi: 10.1016/j.giant.2024.100337

[14] M. I. H. Protyai and A. Bin Rashid, "A comprehensive overview of recent progress in MXene-based polymer composites: Their fabrication processes, advanced applications, and prospects," *Heliyon*, vol. 10, no. 17, p. e37030, 2024, doi: 10.1016/j.heliyon.2024.e37030

[15] E. Sánchez-Díez et al., "Redox flow batteries: Status and perspective towards sustainable stationary energy storage," *J. Power Sources*, vol. 481, 2021, doi: 10.1016/j.jpowsour.2020.228804

[16] L. Huang, Q. Deng, X. Yang, Y. Wang, and W. Dong, "Highly durable perfluorosulfonic acid proton exchange membrane by combining inorganic free radical scavengers and organic antioxidants," *Int. J. Hydrogen Energy*, vol. 69, no. April, pp. 855–862, 2024, doi: 10.1016/j.ijhydene.2024.05.031

[17] K. W. Ahmed, M. J. Jang, M. G. Park, Z. Chen, and M. Fowler, "Effect of components and operating conditions on the performance of PEM electrolyzers: A review," *Electrochem*, vol. 3, no. 4, pp. 581–612, Sep. 2022, doi: 10.3390/electrochem3040040

[18] K. Mensah-Darkwa, D. N. Ampong, E. Agyekum, F. M. de Souza, and R. K. Gupta, "Recent advancements in chalcogenides for electrochemical energy storage applications," *Energies*, vol. 15, no. 11, p. 4052, 2022, doi: 10.3390/en15114052

[19] L. Vilčiauskas and K. D. Kreuer, "Comment on 'mixed grotthuss and vehicle transport mechanism in proton conducting polymers from Ab initio molecular dynamics simulations,'" *Chem. Mater.*, vol. 23, no. 14, pp. 3377–3378, 2011, doi: 10.1021/cm200865v

[20] S. Sreenath, P. S. Nayanthara, M. Krebsz, J. Andrews, and R. K. Nagarale, "Ion exchange membranes: Latest developments toward high-performance vanadium redox flow batteries," *ACS Appl. Energy Mater.*, vol. 7, no. 23, pp. 10846–10876, 2024, doi: 10.1021/acsaem.4c01714

[21] H. Amiri, M. Khosravi, M. Ejeian, and A. Razmjou, "Designing ion-selective membranes for vanadium redox flow batteries," *Adv. Mater. Technol.*, vol. 6, no. 10, pp. 1–16, 2021, doi: 10.1002/admt.202001308

[22] M. T. M. Pham et al., "Prevention of lithium-ion battery thermal runaway using polymer-substrate current collectors," *Cell Reports Phys. Sci.*, vol. 2, no. 3, p. 100360, 2021, doi: 10.1016/j.xcrp.2021.100360

[23] A. Benmehel, S. Chabab, A. L. Do Nascimento Rocha, M. Chepy, and T. Kousksou, "PEM water electrolyzer modeling: Issues and reflections," *Energy Convers. Manag. X*, vol. 24, no. October, p. 100738, 2024, doi: 10.1016/j.ecmx.2024.100738

[24] K. Mensah-Darkwa, D. N. Ampong, E. Agyekum, F. M. de Souza, and R. K. Gupta, "Recent advancements in chalcogenides for electrochemical energy storage applications," *Energies*, vol. 15, no. 11, pp. 1–62, 2022, doi: 10.3390/en15114052

# 2 Classification of Membranes for Electrochemical Devices

## 2.1 INTRODUCTION

Electrochemical devices rely on membranes as selective ion-conducting barriers that enable the controlled ion transport, prevent short circuits, and maintain chemical and mechanical integrity under operational conditions [1]. Depending on their functional requirements, materials, and target ions, electrochemical membranes are classified into different categories with specific structural and operational traits: PEM transports protons ($H^+$) between electrodes, making them suitable for PEMFCs and proton-exchange electrolyzers [2]. Generally, PEMs are made from PFSA or hydrocarbon-based ionomers. As a result, these membranes offer high proton conductivity, low electronic conductivity, low permeability to fuel, good mechanical properties, while maintaining chemical and oxidative stability in acidic environments [3]. Their efficiency depends on hydration levels, the density of sulfonic acid groups, and membrane morphology, as well as mechanical strength to withstand high-pressure operation. AEMs transport $OH^-$ or negatively charged ions, making them suitable for use in alkaline fuel cells, electrolyzers, and redox flow batteries [4]. The membranes have functional groups, such as quaternary ammonium, imidazolium, or phosphonium, attached to the polymer backbone. The most important objectives in designing these membranes include alkaline stability, preventing the formation of carbonate compounds from Carbon dioxide ($CO_2$) absorption, and mechanical durability, preventing swelling and cracking. The most recent advances in AEMs include cross-linking polymers, hybrid materials containing inorganic components, and the use of rigid aromatic backbones to enhance mechanical durability and facilitate ionic transport. Similarly, cation-exchange membranes (CEMs) have electrochemical selectivity, passing positive ions such as $Li^+$, sodium ($Na^+$), and potassium ($K^+$), while excluding negative ions [5]. These materials are critical in lithium-ion, sodium-ion, and redox flow batteries, where ion transport is essential for Coulombic efficiency and the suppression of electrode recombination. Sulfonated polymers or hydrocarbon ionomers are the materials used in the CEMs, and the ion-exchange capability is carefully regulated [6]. On the other hand, non-conductive membranes are permeable and provide physical barriers but are not conductive for ionic conductivity. In the case of batteries and supercapacitors, these membranes are used for the separation of electrodes from each other in order to prevent short circuits and allow for the migration of ions.

DOI: 10.1201/9781003679363-2

The membranes are made from materials like microporous PE, PP, glass fibres, or polymer-ceramic composites [7]. Functionalized and composite membranes are designed for high-performance electrochemical devices, often featuring customized functionalities. Composite membranes combine polymers with inorganic fillers such as silica, zirconia, graphene oxide, and MOFs to improve conductivity, selectivity, mechanical stability, and chemical resistance [8]. Functionalized membranes can have self-healing properties, temperature sensitivity, and redox activity, among others. The use of efficient membrane configurations can improve device efficiency even under harsh or extreme conditions.

Overall, the classification framework presents different types of membranes in a more visual and comprehensive manner, facilitating an improved understanding of material selection, structure, and ion transport behaviour, and is linked to the application-specific requirements. Thus, the detailed characteristics of each of these classes and associated research development, industrial deployment, and performance metrics across a broad range of electrochemical devices are discussed in the following sections.

## 2.2 PROTON-EXCHANGE MEMBRANES

### 2.2.1 Definition and Mechanisms

PEMs refer to the most advanced ionomers, which are specifically made to allow the passage of protons ($H^+$) while blocking electrons and gas molecules, as shown in Figure 2.1. The selective conduction of these types is required in PEMFCs as well as in proton-exchange membrane water electrolyzers (PEMWE) systems, because the efficient proton transport from anode to cathode determines the performance of

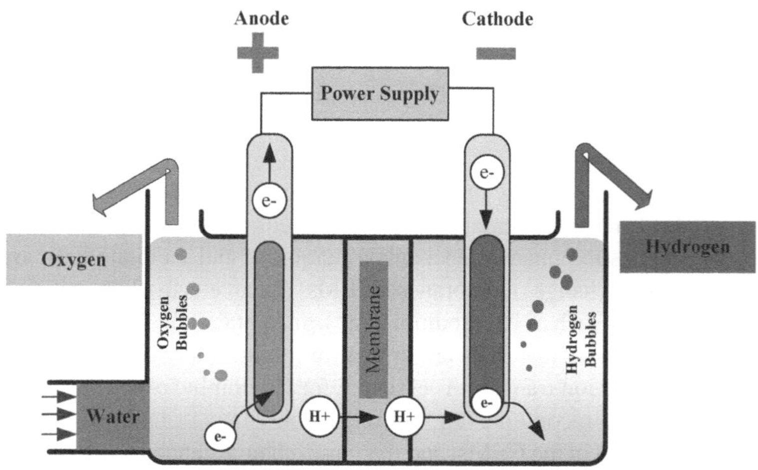


**FIGURE 2.1** Schematic representation of an electrolyzer.

**(Reproduce with permission [3].)**

the entire device [1]. Proton transport in PEMs is generally achieved by two basic mechanisms: (i) the Grotthuss mechanism, in which protons are transported by successive jumps in a hydrogen-bonded chain of water molecules, and (ii) the vehicular mechanism, in which protons are transported in association with diffusing water molecules, for example, in the form of hydronium ions, ($H_3O^+$) [2]. The balance between membrane hydration and morphology determines the ionic conductivity, efficiency, and material durability of the system.

### 2.2.2 Polymeric Membrane Materials

The main materials used in fuel cells are fluorinated ionomers. Nafion is a typical example and serves as the standard for PEMFC. It offers high proton conductivity of about 0.13 S $cm^{-1}$ at 75°C under fully hydrated conditions. It also has strong resistance to oxidative degradation, maintains chemical stability in highly acidic conditions, and has enough mechanical strength to withstand operating stresses. The molecular structure of the Nafion membrane consists of a hydrophobic polytetrafluoroethylene main chain and a perfluoroether branch chain with hydrophilic sulfonic acid groups at the ends. This structure allows the formation of well-connected proton-conducting channels while maintaining structural integrity. New materials are being developed to reduce the cost of fuel cell systems. Hydrocarbon-based materials, such as SPEEK and acid-doped polymers, are considered promising candidates. Their application can also be extended to high-temperature conditions with less humidification, requiring up to 120–180°C. These materials offer high-temperature stability, low material cost, and compatibility with non-precious-metal catalysts. Recent advancements in PEM technology primarily aim to enhance ionic conductivity, mechanical strength, and durability under extreme operating conditions. One such material is sulfonated polyimide nanofiber-reinforced proton-exchange membranes. These membranes are made of a PBI matrix combined with sulfonated polyimide nanofibers. This structure helps form continuous proton-conducting channels and also increases the mechanical resistance to gas permeation [4]. Membranes are very important in high-temperature and low-humidity fuel cell conditions, where Nafion does not retain enough hydration, which results in decreased performance. Composite membranes are an effective way to address challenges simultaneously. For example, PEMs with inorganic materials can provide several advantages, such as enhanced water retention capacity in dry conditions, reduced methanol crossover in direct methanol fuel cells (DMFCs), enhanced thermal stability, and enhanced mechanical properties. These composites also allow for the design of improved microstructures that support efficient ion transport while maintaining dimensional stability. This narrows the gap between laboratory-scale performance and practical device operation. All these improvements indicate a trend towards multifunctional, high-performance PEM designs that integrate high proton conductivity, mechanical reinforcement, and chemical stability. As a result, next-generation fuel cells and electrolyzers equipped with such advanced membrane systems can operate with minimal energy losses, even under demanding industrial conditions and large-scale operations.

## 2.3 ANION-EXCHANGE MEMBRANES

### 2.3.1 Fundamental Principles and Mechanisms

AEMs are ion-conductive polymers that allow the selective movement of anions such as $OH^-$ while restricting the passage of protons, electrons, and other unwanted ions. Because of this selective ion transport, they are widely used in alkaline fuel cells, anion-exchange membrane water electrolyzers, and some redox flow batteries where proper $OH^-$ movement is necessary for electrochemical reactions. In AEMs, ion conduction occurs via Grotthuss-like hopping of $OH^-$ ions through hydrogen-bonded networks and via vehicular transport along with water molecules [5]. The effectiveness of this process depends on the hydration level, the polymer microstructure, and the chemical stability of the cationic head groups.

### 2.3.2 Polymeric Membrane Materials

AEMs are usually prepared by grafting quaternary ammonium or phosphonium groups onto stable polymer backbones, such as polysulfone, polyphenylene oxide (PPO), or poly(aryl ether ketone) (PAEK). These cationic groups provide $OH^-$ conductivity along with electrochemical selectivity. Recent developments include nitrogen-free cationic groups or metal-free alternatives, which are designed to minimize nucleophilic attack and Hofmann elimination, which improves alkaline stability. The selection of the polymer backbone and the position of the functional group are critical; these factors directly affect ionic conductivity, the mechanical strength of the material, and resistance to chemical degradation under strongly alkaline conditions.

### 2.3.3 Emerging Design and Material

The processing of composite materials with innovation and complexity is a major trend in the development of AEMs, which focuses on overcoming the important challenges of low chemical durability, high swelling, and instability in dimensions. An example of a composite material is composed of nanoparticles of silica, graphene oxide sheets, or two-dimensional materials dispersed in a polymer matrix to enhance the membrane's conductivity of $OH^-$ ions, mechanical strength, and dimensional stability during the hydration cycles [6]. In addition, cross-linking techniques and the presence of interpenetrating polymer networks improve the chemical and alkaline resistance of the membranes, thereby extending their operational life in alkaline conditions. These developments aim to achieve an optimal balance of high anion conductivity, mechanical strength, and chemical stability, thereby advancing AEM technology towards practical, cost-effective alternatives to conventional PEMs and alkaline diaphragms in electrochemical devices.

## 2.4 CATION-EXCHANGE MEMBRANES

### 2.4.1 Fundamental Principles and Mechanisms

CEMs are ion-conducting polymers engineered to selectively transport cations such as $Li^+$, $Na^+$, or $K^+$ through fixed anionic functional groups, typically sulfonate ($-SO_3^-$) moieties. [7]. By only allowing the migration of cations and therefore blocking anions and electrons, CEMs perform many crucial functions. These include maintaining an electron balance, which is critical in avoiding cross-contamination and, therefore, allows efficient ion migration in various electrochemical and separation processes. The compatibility and chemical stability of these membranes are crucial factors that directly affect device performance and operational efficiency.

### 2.4.2 Polymeric Membrane Materials

CEMs are usually fabricated in the form of porous membranes for battery applications, which are produced using polymers like PE or PP [8]. The porous membranes not only provide mechanical support and prevent short circuits but also facilitate the flow of electrolyte, a crucial requirement for ionic transport between electrodes. For next-generation lithium-ion and sodium-ion batteries employing solid polymer electrolytes, CEMs are generally made of PVDF, PEO, or their composite derivatives. In most instances, ceramic or nanoparticle fillers are added to improve ionic conductivity and mechanical strength. As a result, the addition of solid-state membranes enables safer battery designs with improved resistance to dendrite formation. These developments have led to next-generation monovalent-ion-selective CEMs, which can be applied for the selective extraction of lithium ions in battery recycling processes. Membrane-based separation for lithium recovery from brine follows a similar technological principle, although the feed sources differ, including spent batteries and natural brines. The techniques used in these membranes include design of pore structures, regulation of functional group density, and cross-linking to create stable network structures. These techniques work in combination to improve selectivity, limit co-ion transport, and improve chemical stability under harsh conditions. The integration of material choice, structural design, and surface modification is an efficient methodology to optimize CEMs according to the demands of high ionic conductivity, mechanical stability, chemical durability, and ion selectivity for energy storage and electrochemical separation applications.

## 2.5 NON-CONDUCTIVE MEMBRANES

### 2.5.1 Fundamental Principles and Mechanisms

Non-conductive membranes serve as either standalone materials that, owing to their porous architecture, primarily function to physically separate the electrodes within electrochemical cells such as LIBs, SIBs, and supercapacitors [9]. Although they do not conduct ions directly, the pores are filled with liquid electrolyte; therefore, the ionic transport occurs between the electrodes, and at the same time, prevents direct electronic, which may cause short circuits. In addition to ensuring safety in the devices, these membranes also provide stability in the functionality of the devices by acting as membranes.

### 2.5.2 Membrane Materials and Key Properties

Non-conductive membranes are normally made of microporous polyolefins, including PE and PP [10]. These materials have good mechanical strength, dimensional stability, and compatibility with battery electrolytes [10]. Moreover, most of these membranes have a thermal shutdown mechanism, whereby the pores close at higher temperatures to stop ionic conduction and prevent further overheating. Another category includes cellulosic membranes, which provide flexibility and high electrolyte wettability and are therefore suitable for thin-film or flexible devices. Gel-polymer membranes are also employed, in which the polymer matrix swells with the electrolyte to impart ionic conductivity while maintaining structural integrity [11]. These membranes can be differentiated based on critical parameters such as porosity, pore size distribution, electrolyte uptake, tensile strength, and thermal resistance. By using appropriate materials for non-conductive membranes, a balance can be achieved among ionic accessibility, mechanical support, and thermal protection.

## 2.6 FUNCTIONALIZED AND COMPOSITE MEMBRANES

Functionalized and composite membranes are referred to as engineered polymer material that involves the use of embedded fillers, nanostructures, or chemical modifications for improved ionic conductivity, selectivity, mechanical properties, and multifunctionality, as shown in Figure 2.2. The membranes are designed to handle the specific requirements of proton, anion, cation transport, or dual-ion conduction, thus enabling applications that utilize such devices to achieve better results [13].

### 2.6.1 PEM Composites

A PEM composite refers to an engineered material formed by introducing inorganic fillers or nanostructures into the polymer, to improve the physical and electrochemical properties. The use of inorganic fillers is one method to enhance the

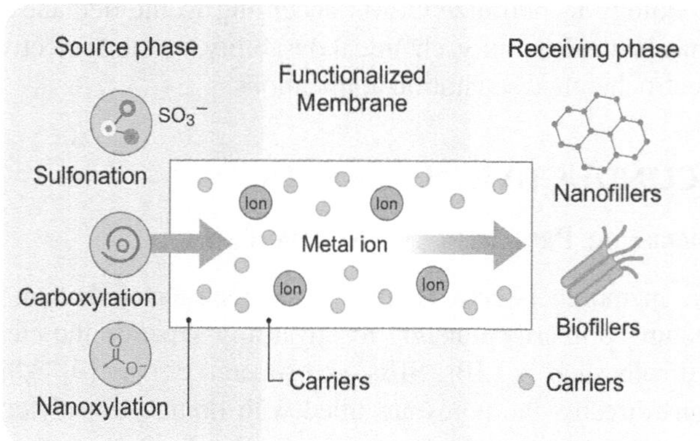


**FIGURE 2.2** Matrix functionalization strategies and their effect on ion transport. **(Reproduced with permission [12].)**

performance of proton-exchange membrane systems. For example, $SiO_2$, zirconium phosphate, or graphene oxide, are typical filler materials incorporated into Nafion or other PFSA matrixes to achieve such goals [14]. These composites could help reduce fuel/methanol crossover, increase dimensional stability, and provide mechanical durability, especially in high-temperature, low-humidity environments. In addition, the fillers help to retain water, which enables proton conductivity to be sustained without swelling and deforming the material. The implementation of such designs enables PEMs to operate stably even after long operational cycles and at high current densities.

### 2.6.2 AEM and CEM Composites

The composite of AEM and CEM can be accomplished through the introduction of specific nanofillers, aiming to optimize the transport of the negatively charged anions or positively charged cations. The incorporation of fillers has been shown to significantly improve the performance of various ion-exchange membranes. Various nanofillers, including graphene oxide, silica, quaternized biopolymers, etc., are effective ways of improving the conductivity of $OH^-/Li^+$ ions, as well as the chemical durability and mechanical strength of anion/cation-exchange membranes. Additionally, these nanofillers help in solving problems like alkaline durability in AEMs and $Li^+$ transport in CEMs, thereby expanding their operational scope [15].

### 2.6.3 Novel Architectures

Recent advances have seen the creation of nanofiber-reinforced polymer matrixes, including polyimide grafted polyethylene oxide (PI-g-PEO) membranes, which have shown promise in improving the performance of lithium-ion batteries [16]. These architectures offer advantages such as continuous ion-conducting channels, increased dendrite suppression, and better mechanical integrity, thereby enabling safe high-rate cycling. Moreover, bipolar and amphoteric membranes, which have both cationic and anionic groups, have the ability to transport ions in both directions, thereby allowing for the creation of reversible or hybrid devices, including fuel cells and redox flow devices [17, 18]. Functional and composite membranes, using the methods of chemical functionalization, the incorporation of nanofillers, and structural engineering, offer an adjustable approach for optimizing the performance of the membranes. The membranes are able to be maximally conductive, selective, stable, and multifunctional, thus paving the way for the development of the next generation of energy technologies. Table 2.1 summarizes the classification of membranes used in electrochemical devices based on their ion-transport function, typical materials, key advantages, and major challenges.

**TABLE 2.1**
**Classification of Membranes for Electrochemical Devices**

| Membrane Type | Ion Transport | Typical Materials | Key Advantages | Major Challenges |
|---|---|---|---|---|
| **PEMs** | Conduct protons ($H^+$) while blocking gases/ electrons | Nafion, SPEEK, PBI, sulfonated polyimides | High proton conductivity, chemical stability, mechanical strength | High cost, dehydration at high T, gas crossover |
| **AEMs** | Conduct hydroxide ions ($OH^-$) or other anions | Quaternary ammonium-grafted polysulfone, PPO, PAEK, imidazolium polymers | Enable alkaline operation, use of non-precious catalysts, cost-effective | Poor alkaline stability, $CO_2$ sensitivity, swelling |
| **CEMs** | Transport cations ($Li^+$, $Na^+$, $K^+$), block anions | Sulfonated polymers, PVDF, PEO composites | Essential for batteries, high selectivity, supports safe ion recovery | Dendrite growth, limited stability in harsh media |
| **Non-Conductive Membranes** | Physical barrier, electrolyte-filled pores allow ion transport | Microporous PE, PP, glass fibre, cellulose-based | Prevents short-circuit, provides thermal shutdown, mechanically robust | Poor wettability, shrinkage at high T, low conductivity |
| **Composite Membranes** | Tailored ion transport ($H^+$, $OH^-$, $Li^+$) with added fillers | Polymer + silica, graphene oxide, zirconia, MOFs | Enhanced conductivity, selectivity, mechanical/ thermal stability | Complexity in fabrication, cost, long-term compatibility |
| **Functionalized Membranes** | Multifunctional: self-healing, thermo-responsive, redox-active | Modified polymers with nanofillers, ionic liquids | Smart response, longer lifetime, multifunctionality | Scalability, reproducibility, stability under harsh cycling |

## 2.7 EMERGING TRENDS AND FUTURE DIRECTIONS

The scope of membrane research is rapidly evolving to meet increasing demands for higher performance, sustainability, and multifunctionality across various electrochemical devices. The development of multi-ion-conducting and bipolar membranes that combine both $H^+$ and $OH^-$ selective pathways within a single membrane marks one of the major directions in this area [19]. These structures facilitate reversible operation, thus supporting hybrid systems, for example, reversible fuel cells or battery-fuel cell combinations, while providing new prospects for the design of flexible energy conversion and storage. Nanofiber architectures represent one of the most significant advancements in the membrane field [20]. Incorporating electrospun or self-assembled nanofibers into membrane materials has been shown to create continuous channels that are highly conductive, while at the same time allowing the achievement of sufficient mechanical properties [21]. Biomass-based green materials used in the fabrication of ionomer materials have been shown to offer an effective solution in terms of increasing sustainability in membrane materials. This is achieved by allowing the achievement of similar ionic conductivity properties as those of traditional materials, while at the same time reducing the use of fluoropolymers. Membranes are being fundamentally re-designed using novel fabrication techniques. The different methods have different capabilities, such as the precision control of pore size, the distribution of the functional groups, thickness, and smoothness of the surface, to name a few. These features enable the design of customizable ionic pathways, enhanced selective permeability, and improved mechanical strength, while meeting the performance requirements of advanced fuel cells, electrolyzers, and battery technologies. Together, these trends signal a future full of membranes that are highly efficient, multifunctional, and sustainable, crossing over effortlessly from the laboratory to large-scale applications that are industrially viable as energy solutions.

## 2.8 CONCLUSION

In this section, the key features, applications, and recent innovations of electrochemical membranes are summarized, emphasizing the interrelationships among ion transport, material selection, and functional enhancements. PEMs are responsible for the transport of $H^+$ ions and serve as a key component in fuel cells and PEM electrolyzers. Materials such as Nafion, SPEEK, and PBI are benchmark materials that offer excellent proton conductivity and chemical stability. These properties enable stable operation under high-temperature and low-humidity conditions. AEMs are the ones that carry hydroxide ions, and, consequently, they are used in alkaline fuel cells and, in the case of an alkaline solution, water-shredding is carried out by means of AEM. Polymers having supported functional groups such as quaternary ammonium or phosphonium on a stable polymer backbone provide the characteristics of chemical stability and structural integrity sufficient to withstand the attack of an alkaline environment. Progresses like composite fillers, crosslinked networks, and advanced head groups do enhance those features of the membranes that include ion conductivity, durability, and resistance to carbonate formation. Polymeric materials like polyolefins, PVDF, PEO, and their composites are what provide a battery with the mechanical integrity and electrochemical stability that it requires. The incorporation of monovalent-ion selectivity along with robust solid polymer electrolyte systems has been pivotal in achieving safer, high-performance battery technologies and in facilitating the advancement of lithium recycling processes. Composite and Multifunctional Membranes have been developed to integrate $H^+$, $OH^-$, and metal-ion conduction for multi-device platforms. By the use of MOFs, graphene oxide, nanofillers, and hybrid polymers, membranes combine several advantages of high selectivity, high mechanical strength, the possibility of ionic conduction, and multifunctionality, so that they are suitable as the next generation of energy conversion and storage systems. Membrane development has progressively evolved from simple selective materials to multifunctional, durable, and application-specific systems, reflecting the continuous advancement in material design and engineering. The design of these membranes is a mesh of material chemistry, structure, and device performance. Such membranes play a fundamental role in enabling highly efficient, safe, and scalable electrochemical technologies.

## REFERENCES


[1] J. Mao, Z. Li, J. Xuan, X. Du, M. Ni, and L. Xing, "A review of control strategies for proton exchange membrane (PEM) fuel cells and water electrolysers: From automation to autonomy," *Energy AI*, vol. 17, no. July, p. 100406, 2024, doi: 10.1016/j.egyai.2024.100406

[2] I. Popov et al., "Search for a Grotthuss mechanism through the observation of proton transfer," *Commun. Chem.*, vol. 6, no. 1, 2023, doi: 10.1038/s42004-023-00878-6

[3] H. Marefatjouikilevaee, F. Auger, J. C. Olivier. Static and dynamic electrical models of proton exchange membrane electrolysers: A comprehensive review. *Energies*. vol. 16, no. 18, p. 6503, 2023 Sep 9, doi: 10.3390/en16186503

[4] K. Khoiruddin, C. Supriyadi, A. Nandar, S. Kawi, T. M. Lim, and I. Gede, "Advanced proton exchange membranes for high- efficiency fuel cells: Material innovations and durability optimization," *Energy Mater.* vol. 5, p. 500141, 2025, doi: 10.20517/energymater.2025.62

[5] Z. Wang et al., "Water-mediated ion transport in an anion exchange membrane," *Nat. Commun.*, vol. 16, no. 1, pp. 1–10, 2025, doi: 10.1038/s41467-024-55621-z

[6] T. Shahnaz et al., "Graphene-based nanoarchitecture as a potent cushioning/filler in polymer composites and their applications," *J. Mater. Res. Technol.*, vol. 28, no. November 2023, pp. 2671–2698, 2024, doi: 10.1016/j.jmrt.2023.12.108

[7] Y. Kakihana, N. A. Hashim, T. Mizuno, M. Anno, and M. Higa, "Ionic transport properties of cation-exchange membranes prepared from poly(Vinyl alcohol-b-sodium styrene sulfonate)," *Membranes (Basel).*, vol. 11, no. 6, 2021, doi: 10.3390/membranes11060452

[8] M. T. Tsehaye et al., "Membranes for zinc-air batteries: Recent progress, challenges and perspectives," *J. Power Sources*, vol. 475, no. August, p. 228689, 2020, doi: 10.1016/j.jpowsour.2020.228689

[9] L. Li and Y. Duan, "Engineering polymer-based porous membrane for sustainable lithium-ion battery separators," *Polymers (Basel).*, vol. 15, no. 18, 2023, doi: 10.3390/polym15183690

[10] A. A. Heidari and H. Mahdavi, "Recent development of polyolefin-based microporous separators for li–ion batteries: A review," *Chem. Rec.*, vol. 20, no. 6, pp. 570–595, 2020, doi: 10.1002/tcr.201900054

[11] Z. Chen et al., "Advanced cellulose-based materials for flexible energy storage systems," *Resour. Chem. Mater.*, vol. 4, no. 3, p. 100120, 2025, doi: 10.1016/j.recm.2025.100120

[12] A. Nowik-Zaj, "Recent developments in polymer inclusion membranes: Advances in selectivity, structural integrity, environmental applications and sustainable fabrication," *Membranes*. vol. 15, no. 8, p. 249, 2025, doi: 10.3390/membranes15080249

[13] G. B. Appetecchi, "Composite electrolyte & electrode membranes for electrochemical energy storage & conversion devices," *Membranes (Basel).*, vol. 10, no. 11, p. 359, Nov. 2020, doi: 10.3390/membranes10110359

[14] A. Islam et al., "Advances in filler-crosslinked membranes for hydrogen fuel cells in sustainable energy generation," *Int. J. Hydrogen Energy*, vol. 140, no. May, pp. 745–776, 2025, doi: 10.1016/j.ijhydene.2025.05.197

[15] S. M. S. Al-Mufti, A. Almontasser, S. J. A. Rizvi, and R. K. Arya, "Applications of nanofillers in proton exchange membrane fuel cells," in *Handbook of Nanofillers*, Singapore: Springer Nature Singapore, 2024, pp. 1–31. doi: 10.1007/978-981-99-3516-1_91-1
[16] H. Kawakami, "Development of composite electrolyte membranes with functional polymer nanofiber frameworks," *Polym. J.*, vol. 57, no. 6, pp. 623–633, Jun. 2025, doi: 10.1038/s41428-024-01007-3
[17] I. Belhaj, M. Faria, B. Šljukić, V. Geraldes, and D. M. F. Santos, "Bipolar membranes for direct borohydride fuel cells—a review," *Membranes (Basel).*, vol. 13, no. 8, p. 730, Aug. 2023, doi: 10.3390/membranes13080730
[18] D. Han and S. Shanmugam, "Active material crossover suppression with bi-ionic transportability by an amphoteric membrane for Zinc–Bromine redox flow battery," *J. Power Sources*, vol. 540, p. 231637, Aug. 2022, doi: 10.1016/j.jpowsour.2022.231637
[19] J. C. Bui, I. Digdaya, C. Xiang, A. T. Bell, and A. Z. Weber, "Understanding multi-ion transport mechanisms in bipolar membranes," *ACS Appl. Mater. Interfaces*, vol. 12, no. 47, pp. 52509–52526, Nov. 2020, doi: 10.1021/acsami.0c12686
[20] P. Kallem, N. Yanar, and H. Choi, "Nanofiber-based proton exchange membranes: Development of aligned electrospun nanofibers for polymer electrolyte fuel cell applications," *ACS Sustain. Chem. Eng.*, vol. 7, no. 2, pp. 1808–1825, Jan. 2019, doi: 10.1021/acssuschemeng.8b03601
[21] S. N. Banitaba and A. Ehrmann, "Application of electrospun nanofibers for fabrication of versatile and highly efficient electrochemical devices: A review," *Polymers (Basel).*, vol. 13, no. 11, p. 1741, May 2021, doi: 10.3390/polym13111741

# 3 Preparation and Fabrication of Membranes

## 3.1 INTRODUCTION

The performance of membranes in electrochemical devices such as Lithium-Ion Battery (LIBs), fuel cells, electrolysers, and redox flow batteries is crucially dependent on the structural and chemical parameters such as morphology, porosity, thickness, crystallinity, and composition [1]. These parameters affect the ionic conductivity and the wettability of the electrolyte, as well as the mechanical stability, dimensional stability, thermal shrinkage, and chemical stability factors that affect the performance of electrochemical devices under harsh operating conditions [2]. However, it is critical to optimise these parameters to obtain optimal performance, safety, and energy efficiency with the desired lifespan [3]. The methods employed in membrane fabrication include phase separation, mechanical stretching, electrohydrodynamic fibre formation methods, and additive manufacturing, which enable the formation of unique membrane structures through the control of phase separation, pores, fibre orientation, and filler content [4]. Conventional methods such as solution casting, phase inversion casting, and extrusion have remained the main methods in the industrial production process because they are relatively simple, scalable, and inexpensive; however, they lack the capability to engineer nanoscale features and multifunctionality [5]. New methods such as electrospinning, melt electrospinning, and 3D printing can be complex; however, they may not be necessary for all electrochemical devices [6]. Methods such as layer-by-Layer (LbL) assembly and other advanced coating methods can be used to control the chemistry at the nanoscale and can be used to design selective ion transport pathways [7]. This chapter will present a comprehensive overview of the conventional as well as innovative techniques for fabricating membranes, discussing the advantages and challenges of the techniques, while highlighting the recent developments in membranes and membrane technology for electrochemical applications. It will cover issues of the relationship between process, structure, and properties, including ion transport, stability, and safety, as well as new hybrid fabrication techniques, and finally discuss future directions for rationalising membranes in energy storage and conversion technologies, providing insights for designing next-generation membranes with performance and safety criteria of existing systems.

## 3.2 SOLUTION CASTING AND PHASE INVERSION

Phase inversion is a demixing process in which a homogeneous solution of polymers gradually changes from a liquid to a solid under controlled conditions. This process


DOI: 10.1201/9781003679363-3

of phase inversion is considered to be one of the most commonly used methods of preparing polymeric porous membranes with good structures. Phase inversion works by mixing polymer and a suitable solvent, forcing them to separate into polymer-rich and polymer-poor phases. The process is governed by thermodynamics, specifically the shift from a stable one-phase region to an unstable two-phase region on a ternary phase diagram. Depending on how the solvent is removed and how the non-solvent is introduced, phase inversion can be divided into four types: non-solvent-induced phase separation (NIPS), thermally induced phase separation (TIPS), vapour-induced phase separation (VIPS), and evaporation-induced phase separation (EIPS).

NIPS is a widely used membrane production technique in which liquid-liquid demixing occurs. In this case, the rate of diffusion between the solvent and the non-solvent determines the morphology of the final membrane, which can be dense, porous, spongy, symmetric, or asymmetric. In TIPS membranes, various parameters affect the membranes, such as the exclusion or evaporation of the diluent from the surface of the membrane, resulting in a higher concentration of polymers on the surface and rapid cooling of the solution. In the VIPS process, the casting solution is subjected to air or steam for a certain period of time before it is immersed in a coagulation bath for phase separation and the formation of the membrane [8]. This period of time is called the pre-evaporation time ($V_T$). In EIPS, a homogeneous solution is prepared by dissolving the polymer with the solvent and non-solvent mixture, with the solvent being more volatile than the non-solvent. This results in phase separation and the process of demixing the polymer-solvent-non-solvent system, leading to the porous films (Figure 3.1) [10].

One of the oldest and most flexible techniques for fabricating polymer membranes is by blending solution casting with phase inversion. There are several reasons for choosing this technique, which are as follows: simple and flexible for use in different

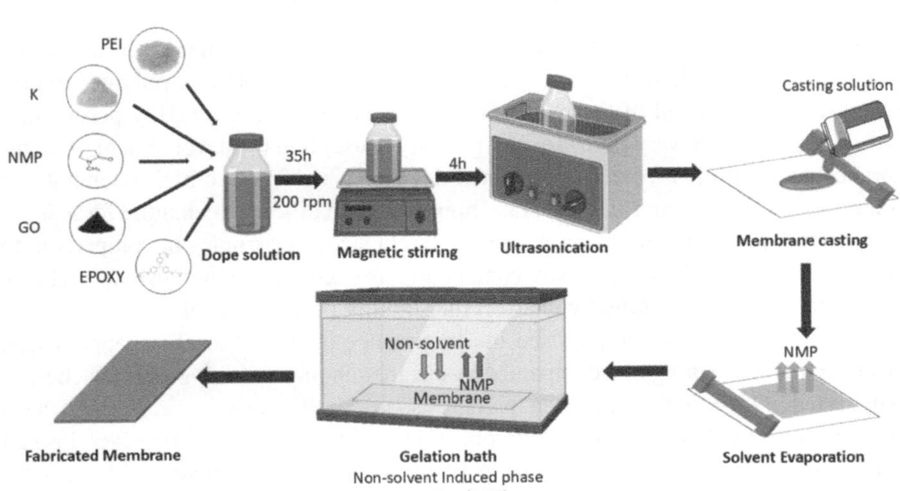


**FIGURE 3.1** Schematic representation of phase-inversion techniques.

**(Reproduce with permission [9].)**

applications, scalability, along with the suitability of this technique for different kinds of polymers, and the ability of solution casting to produce thin, defect-free films with tunable properties for different kinds of electrochemical devices, such as LIB membranes, PEMs, and AEMs [11, 12]. The fundamental concept is to create a phase separation in a cast polymer solution, creating enriched and lean phases, which can be either polymer-rich or lean. Modifications of this approach differ in how the phase separation is initiated. The approach for making a membrane directly impacts the pore size distribution, interconnectedness, and morphology, which in turn affect electrolyte uptake, conductivity, and selectivity. However, after several years of use in industrial applications, classical phase-inversion techniques are having difficulty in achieving nanoscale pore size uniformity and addressing issues of solvent detachment, which are very important for long-term stability in electrochemical environments. Recently, new hybrid techniques have been investigated, such as incorporating inorganic materials like $SiO_2$, Titanium dioxide ($TiO_2$), and $Al_2O_3$, using block copolymers to create ordered pore structures, and green chemistry techniques [13].

Other techniques involve hybridising TIPS and VIPS, followed by stretching or coating. The mechanical properties, along with the performance, of the membranes developed by this method may lead to their use in structural applications, apart from their use in primary separation functions. To conclude, phase inversion is a widely used membrane fabrication method, and advancements in this technique primarily focus on improving membrane properties during fabrication rather than through post-fabrication modifications. Improvements in this method in the future may be focused on controlling the pore structure at multiple levels, using solvents in the development process, or making composite phase-inverted membranes with high ionic conductivity, dimensional stability, and electrochemical stability in extreme environments.

## 3.3 SOLVENT-CASTING PARTICULATE LEACHING

The particulate leaching technique is an extension of the convenient solvent-casting technique, in which sacrificial templates are incorporated into a polymer matrix (Figure 3.2). Removal of these sacrificial templates results in a highly porous network. The ability to vary pore size and porosity is useful in the development of membranes and scaffolds for electrochemical devices, which require high retentivity of electrolyte solutions and stability. Leaching is a more efficient technique for achieving desired pore sizes by incorporating different sizes of particles in comparison to phase inversion, although porosity content affects strength and dimensional stability [15]. However, pore volumes can affect the stability of the material.

Recent studies have emphasised the importance of hybrid leaching approaches, where particle leaching is accompanied by electrospinning and sol–gel approaches, as suggested in [16]. In this approach, particles such as salts, sugars, starches, and crystals are mixed with the polymer solution, similar to solvent casting; however, instead of casting the film and then adding the particles, the particles are mixed with the polymers before casting the film. After casting the film and evaporating the solvent, the particles are leached using water or a chemical solvent, as needed. The material has interconnected pores and contains particles, which is beneficial for electrochemical applications since the pore diameters can be controlled using different particles and

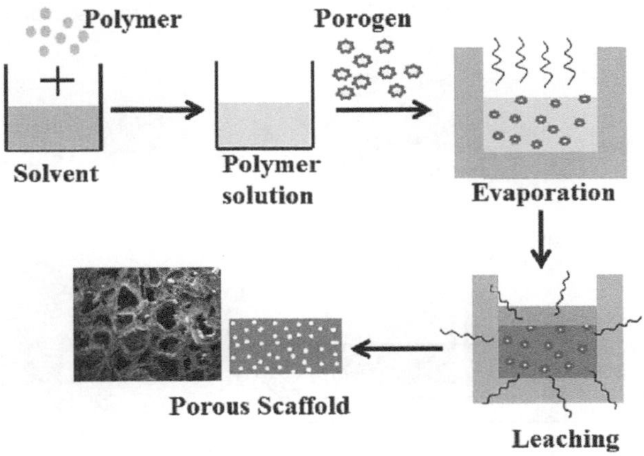


**FIGURE 3.2** Schematic representation of solvent casting and particulate leaching technique. **(Reproduce with permission [14].)**

their amounts. The versatility of this technique also makes it highly suitable for use in making membranes for use in applications such as lithium-ion batteries, solid polymer electrolytes, and composite scaffolds/membranes that need high ionic conductivity without compromising other properties. However, it should be noted that this technique also has some limitations. Excessive porosity due to too many or too large particles may lead to weakening of the material or instability of its shape. Impurities may also be introduced by the sacrificial particles, which may be problematic for their interaction with liquid electrolytes. However, to overcome these problems, hybrid leaching techniques have been investigated for use in making membranes for different applications. An example of this is the use of salt leaching/electrospinning, which is used to produce layered membranes for use in high ionic conductivity applications.

Another method is sol–gel leaching assisted by size control, which creates areas of low aspect ratio wires rather than interconnects due to localised leaching activity. A strong, thermally stable composite membrane is achieved [17]. Trends in current technology indicate that biodegradable sacrificial materials are becoming increasingly popular as a green technology for membranes. The idea of particulate leaching is becoming increasingly popular in the design of advanced membranes with macro-, micro-, and nanoscale pores for efficient ionic conduction without compromising strength. The idea of functional fillers and green technology for membranes prepared by particulate leaching is likely to be increasingly popular in the near future for high-performance green membranes for battery and fuel cell applications.

## 3.4 EXTRUSION, STRETCHING, AND TRACK ETCHING

Extrusion methods, followed by stretching or track etching, provide the basis for commercial-scale polyolefin membrane production worldwide, and the technology is the most popular method for LIB applications. The mechanical and thermal treatments exploit the semicrystalline character of polymers such as PE and PP to create

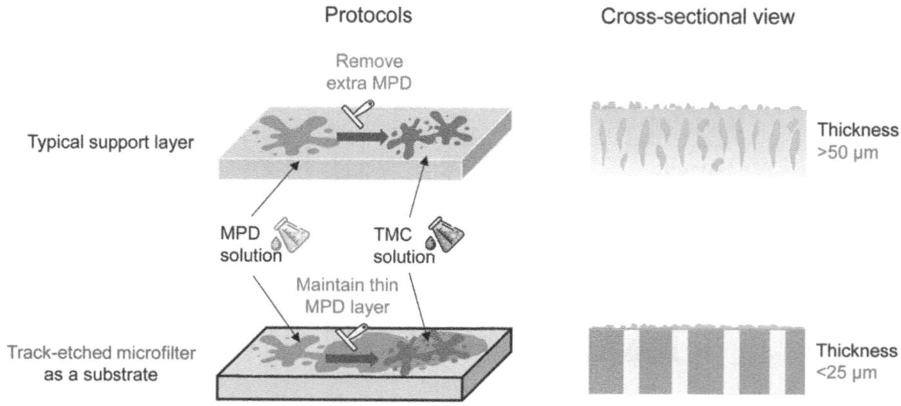


**FIGURE 3.3** Schematic representation of track-etched membrane preparation. **(Reproduce with permission [18].)**

microporosity, as illustrated in Figure 3.3. The advantages of microporous membranes can be attributed to the material properties, which include a high strength-to-weight ratio, low cost, and the capability to produce membranes on a large scale for LIBs [19]. In dry processing, also referred to as mechanical stretching, the polyolefin membrane is first extruded and then annealed and stretched uniaxially or biaxially [20]. The stretching process involves cavitation in the polymer, which occurs at the crystalline-amorphous boundaries, and the cavitation results in the formation of a porous polymer matrix with fibrils. The method can produce at least two types of pores that are elongated and interconnected and can provide uniform paths for ionic transport with sufficient mechanical integrity. On the other hand, the wet method incorporates pore-forming agents that are subsequently removed. Apart from enhancing the pore-stretching effect, it also generates isotropy in pore size.

The track etch method is another technique for controlling pores in membranes. The polymer is irradiated with high-energy ions to produce latent tracks of damage [21]. These tracks are then developed to create highly uniform cylindrical pores of controlled diameters in the range of tens of nanometres to several Microns in size. These membranes are particularly useful when uniformity in ion transport is required, but their high cost and low production rate limit their use on a large scale [22]. Moreover, whereas the stretching or track etching technique is often employed for polyolefin membranes, these membranes have some disadvantages of their own. Being hydrophobic in nature, they are not easily wettable with liquid electrolytes, thereby showing inferior performance in terms of absorption and ionic conductivity compared to other membranes. Additionally, they are known to shrink upon heating, thereby showing inferior safety characteristics in terms of thermal runaway and internal shorts. In order to improve the characteristics of these membranes, surface modification and hybridisation approaches have been studied, including the technique of integrated modification. In this regard, surface modification has been performed using ceramic particles ($Al_2O_3$, $SiO_2$, and $TiO_2$), polymer layers (PVDF, Polyimide (PI), or electrolyte-philic copolymers), and flame retardants [23].

Multilayer composites, such as PP/PE/PP trilayers or polyolefin–ceramic hybrids, are already commercially available and have shown promise in improving electrolyte compatibility, acting as shutdown or fire-resistant layers, respectively. Furthermore, the use of nanoparticles incorporated throughout the extrusion or stretching process can also be effective in improving thermal properties and ionic conductivity. It is also expected that industrial development will continue to focus on scaling up the process, such as extrusion and track etching, with functional hybridisation being a key competitive strategy. Future polyolefin membrane technology is also expected to involve multi-scale structures or designs developed throughout the process, such as ceramic/polymer coatings and flame-retardant features. An ability to scale up a polymer battery membrane process, such as extrusion, and incorporate coatings or hybridisation is crucial in meeting requirements for safety and performance characteristics over a range of temperatures, particularly for high-energy density LIBs and beyond.

## 3.5 ELECTROSPINNING AND MELT ELECTROSPINNING

Electrospinning has been identified as a highly effective and flexible technology for the fabrication of ultrathin and nanofibrous membranes possessing unique structural and functional characteristics. It is carried out by extruding ultra-fine polymer fibres through the application of electrohydrodynamic forces in the presence of a strong electric field, as depicted in Figure 3.4. This fabrication route is entirely dissimilar from the conventional routes followed for the fabrication of membranes based on thermodynamic and diffusion principles. The majority of the membranes are composed of non-woven mats consisting of overlapping and interconnected fibres having

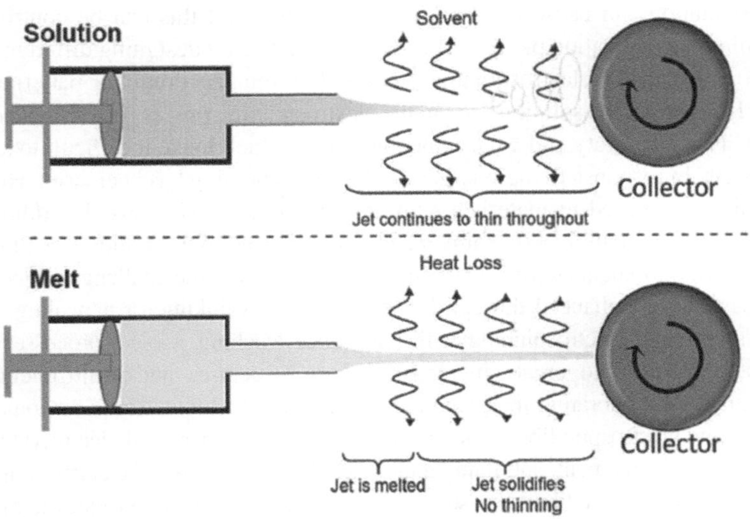


**FIGURE 3.4** Schematic illustration demonstrating fundamental differences between solution and melt electrospinning.

**(Reproduced with permission [24].)**

extremely high porosities (60–90%), large surface areas, and variable orientations and pore structures [25]. The high surface area and pore structure are highly advantageous for electrochemical applications that demand rapid ionic transport rates, high electrolyte uptake capacity, and mechanical flexibility. Solution electrospinning involves the extrusion of the polymer solution through the needle tip, with the surface tension being broken by the application of a high voltage. This generates a charged jet that eventually forms into a very fine fibre as it is stretched and undergoes whipping instabilities, with the diameter being significantly smaller than the initial jet diameter. As the solvent evaporates, the nanofibers are deposited on the collector. The fibre morphology, orientation, and porosity depend on parameters such as the polymer concentration, applied voltage, solution viscosity, and collector shape.

Melt electrospinning and its innovative derivative, melt electrospinning writing, do not require the application of solvents because they melt the polymer and directly extrude it through the spinneret tip. This is an advancement over the conventional method because it can produce ordered nanofibers, which is not possible with the random arrangement of solution electrospinning. The application in the field of electrochemistry is being researched and is promising. For example, in the application of LIBs, the electrospun membranes prepared from PVDF, PAN, PI, and their composites show superior electrolyte wettability, higher ionic conductivity, and enhanced thermal stability compared to the traditional polyolefin membranes used [26]. In solid-state batteries, nanofibers are used to create scaffolds for polymer electrolytes, which are useful in achieving the dispersion of ceramic particles and ionic conduction. Moreover, in fuel cells, nanofibers are useful in acting as catalyst supports, owing to their high surface area for catalyst utilisation and ease of diffusion of gases. There are also some disadvantages associated with the use of electrospinning technology. The weak inter-fibre binding of electrospun membranes leads to poor mechanical strength. Random fibre orientation can cause non-uniform ionic paths, and this can be controlled by controlling the deposition pattern. This can be achieved by designing different collector types used in the electrospinning process, like rotating drums or patterned electrodes. In addition, the solution used in the electrospinning process is a volatile organic solvent. This has safety and environmental issues, which make it difficult to scale up the process. In response to these issues, melt electrospinning has been used. However, the method is limited to materials with appropriate melt viscosity. In addition, the method is not compatible with some materials compared to the solution method.

Recent developments have been focused on overcoming the challenges. Mechanical properties can be enhanced through hot pressing, cross-linking, or annealing. Hybrid methods combine electrospinning with particulate leaching, sol-gel processes, or in-situ polymerisation to create hierarchical pore structures and multifunctionality. Additionally, incorporating inorganic nanoparticles, 2D nanomaterials, or ionic liquid additives into electrospun fibres imparts properties such as increased electrolyte affinity, mechanical reinforcement, and flame retardancy. Hybrid core–shell electrospinning can produce multifunctional fibres that serve as membranes capable of intrinsic thermal shutdown or acting as reservoirs of electrolytes. Electrospinning has the potential of connecting the nanoscale decision-making with macro-scale device usability and may be one of the best value-added production methods for niche markets such as membranes and advanced membranes that will shift as the demand for electrochemical devices evolves.

## 3.6 WET AND DRY SPINNING

Spinning methods, which were initially developed in the textile industry, are still used in the production of membranes and fibrous materials used in electrochemical applications. In both wet spinning and dry spinning, a solution of a polymer is forced through a spinneret, but they differ in the process of solidification.

1. **Wet Spinning**: The solution of polymers is then extruded into a bath of coagulation, which contains a non-solvent. Solidification takes place through a process of phase separation in the coagulation bath, where the solvent diffuses out of the solution of polymers, while the non-solvent diffuses into the solution of polymers, leading to the solidification of polymers in continuous filaments.
2. **Dry Spinning**: The polymer solution is extruded in a heated chamber, and the solvent evaporates, leaving behind the polymer fibres.

These techniques can be used to produce continuous fibres with diameters ranging from nanometres to micrometres, depending on the spinning conditions, concentration, and properties of the solvent. These spinning techniques are specifically applicable in electrochemical systems where high-strength fibres or fibrous membrane structures are required. Polyacrylonitrile (PAN) is the most common precursor used for electrode fibres that will be transformed into conductive carbon fibres through the stabilisation and carbonisation process, making it an essential material in the electrode and current collector of energy storage devices. However, it is worth mentioning that wet spinning of PAN and its derivatives has allowed the development of fibrous membranes and scaffolds with enhanced mechanical properties.

In the same way, other conductive polymers such as polyaniline (PANI) and polypyrrole (PPy), as well as nanocomposite fibres containing carbon nanotubes (CNT) and graphene, have been prepared using the spinning method for the development of electrodes, solid polymer electrolytes, and membranes [27]. Recent developments have focused on improving the performance and free-standing PAN fibres that have been doped with CNTs and graphene, which have been prepared through wet spinning, and show enhanced electrical conductivity and tensile strength, making them suitable for structural and flexible battery applications. Wet-spun hybrid fibres with in-situ ionic liquid domains have also been tested and evaluated with respect to ionic connectivity. Dry-spun fibrous mats with the help of coating and cross-linking methods can improve chemical stability and resistance. In addition, some researchers have also focused on coaxial spinning, which can improve the multifunctionality of the membrane through the incorporation of a core-shell structure that can show enhanced functionality, self-healing, flame resistance, and ionic and electronic conductivity [28]. The future focus of membrane spinning could be the incorporation of green chemistry with the help of the use of green solvents and green coagulants. Prospects of spinning-derived membranes could include the application of green coagulants and solvents in the spinning process to ensure the minimisation of environmental effects. Multicomponent fibre materials, which comprise the spinning of polymers, nanoparticles, and ionic conductors into a single fibre, could be extended to include the application of post-processing methods such as carbonisation, electroplating, and coating to generate multifunctional membranes and electrodes. The expansion of coaxial and bicomponent spinning methods could be employed to generate membranes with hierarchical structures and functions. Wet and dry spinning methods are crucial in the generation of advanced electrochemical devices due to the mechanical properties that can be achieved with the spinning process.

## 3.7 3D PRINTING

Additive manufacturing, more commonly known as 3D printing, has recently started to revolutionise the membrane fabrication domain. Conventional membrane fabrication methods involve casting, stretching, or spinning, whereas 3D printing is different from the conventional methods in the sense that it allows for the fabrication of membranes based on digital designs. This allows for accurate control of the membrane characteristics in electrochemical applications, where it is essential to control the thickness of the membrane, the pore distribution, and the ionic paths individually and precisely in order to attain a balance of performance, safety, and durability. Several 3D printing methods are under investigation for membrane fabrication:

1. **Fused Deposition Modelling**: As per thermoplastic filaments such as PVDF or PLA, the filament is melted and extruded in an LbL manner. This is a cost-effective method, which is easily scalable but is limited by the nozzle size, which affects print resolution.
2. **Stereolithography (SLA) and Digital Light Processing**: Light patterns enable the high-fidelity polymerisation of UV-curable resins with microscale resolution and smoothness.
3. **Ink-jet Printing and Slot-Die Coating**: Although these techniques are not strictly three-dimensional printing techniques, they can be used for patterned deposition of thin functional films, possibly in conjunction with more conventional printing techniques.

In fuel cells, incorporating hierarchical channels in 3D-printed membranes and gas diffusion layers enhances the transport of reactants and water management. In redox flow batteries, structured pores in SLA-fabricated ion-selective membranes reduce the crossover of active species. In electrolysers, 3D-printed membranes and electrodes optimise gas evolution pathways and mass transport. Furthermore, recent developments include hybrid 3D printing techniques that combine additive manufacturing with methods like electrospinning or sol–gel methods to create multi-scale architectures. Researchers are also exploring the addition of nanofillers such as ceramics, MOFs, CNTs, and graphene into printable inks for improved conductivity, mechanical strength, and chemical stability [29]. Additionally, bio-inspired printing techniques are being developed to create gradient structures and asymmetric membranes that mimic natural ion channels. Printable inks that can be used with a wider spectrum of polymers and functional fillers have also been reported. Work is underway to scale up multi-nozzle and continuous printing systems to bridge the gap between laboratory prototypes and commercialisation.

The application of machine learning and computational design is also being investigated to improve membrane structures for targeted electrochemical applications. The design and synthesis of multifunctional membranes with ionic conductivity, catalytic, and mechanical properties without compromising printing accuracy is also being targeted. Additive manufacturing has the advantages of digital accuracy and design freedom in membrane design. Although there are challenges related to scalability and performance metrics, the convergence of 3D printing, materials chemistry, and computational design makes this approach promising for the next generation of electrochemical devices.

## 3.8 LAYER-BY-LAYER ASSEMBLY AND COATINGS

The LbL assembly method is the most accurate and flexible method for modifying the surface of the membrane and fabricating ultrathin films. LbL assembly is a method in which two materials having opposite charges in the form of polyelectrolytes, nanoparticles, and molecules are deposited in layers in a step-by-step manner through the action of attractive forces such as electrostatic forces, covalent forces, and H-bonding [30]. It has the advantage of adding only a few nanometres in thickness in each bilayer. This method allows the surface engineering of the membrane, which is very different from the other methods and allows the modification of the existing membranes without disturbing the mechanical properties. In the field of electrochemistry, the LbL method has the advantage of solving the problems associated with the use of commercial membranes. For example, in lithium-ion batteries, the electrochemical stability and hydrophobic nature of the membranes composed of polyolefin are the major problems. They are usually wetted very poorly by the electrolytes.

Likewise, in redox flow batteries, the use of LbL films composed of ion-selective polyelectrolytes and nanomaterials has been reported as an effective method for minimising the crossover of vanadium ions in the battery, hence improving coulombic efficiency. In the field of fuel cells, multifunctional LbL films consisting of proton-conducting polymers and catalytic nanoparticles were reported as protective films for improving conductivity and durability. Although the LbL method has many advantages over other methods, such as its modularity and the capacity for the integration of multifunctional layers (hydrophilicity, flame resistance, catalytic properties, and ion selectivity), its nanoscale thickness control, and the capacity for the integration of almost any charged and functionalizable material (excellent level of compatibility), the method has its own disadvantages. One disadvantage is the time required for the sequential deposition of the LbL films. Moreover, the mechanical properties of the LbL films are weak and delaminate unless the two surfaces are cross-linked or chemically bonded over a longer time. The disadvantages are currently being addressed through the use of spray-assisted LbL, roll-to-roll LbL, and ink-jet-LbL.

Furthermore, small amounts of nanomaterials like graphene oxide, MXenes, MOFs, and nanoparticles can be used to enhance the mechanical strength, ionic channels, and catalysts of the membranes. Another aspect being explored is the hybrid approach of electrospinning and LbL. Although LbL is currently being explored as a tool for surface modification, it is expected that the technique will soon become mainstream and pre-standardised for the manufacturing of membranes. It is expected that the future of LbL will involve the following aspects: the development of automated solutions for the production of membranes using LbL, the incorporation of smart materials to produce adaptive membranes, and the use of bio-inspired approaches that utilise the properties of biological membranes like ion channels [31]. LbL-assembled membranes have the capability to be engineered at the nanoscale while still providing the membranes with the capability of being multifunctional. Therefore, LbL-assembled membranes can provide the required performance and long lifespan for the next generation of electrochemical energy devices.

## 3.9 COMPOSITE MEMBRANE FABRICATION

While the technical difficulties posed by the limitations of single-component polymer membranes are major challenges, the fabrication of composite polymer membranes has been identified as the most popular approach in addressing the limitations of polymer membranes. The addition of other polymers, such as inorganic nanofillers, secondary polymers, mixed polymers, and segmented polymers, improves the performance characteristics, such as ionic conductivity, mechanical properties, thermal properties, and selectivity in composite polymer membranes, as shown in Figure 3.5. Composite polymer membranes are particularly useful in the fabrication of challenging electrochemical devices such as LIBs, SSEs, FCs, redox batteries, and so forth. There are various methodologies in the fabrication of composite polymer membranes.

1. **Physical Blending** In this traditional approach, inorganic fillers like $SiO_2$, $TiO_2$, $Al_2O_3$, Zirconium dioxide ($ZrO_2$), zeolites, montmorillonite, graphene oxide, and MOFs are dispersed in a polymer matrix like PVDF, PAN, and SPEEK. These fillers can modify the crystalline nature of the polymer, which can, in turn, improve the segmental mobility and ionic conductivity. In addition, rigid fillers can improve the mechanical and thermal durability of the membrane.
2. **Sol-Gel Processing** In-situ synthesis of inorganic nanoparticles in a polymer solution through sol–gel technology enables the homogeneous dispersion and strong interfacial adhesion of the nanoparticles. This method is effective in incorporating silica, zirconia, and titania nanoparticles, which improve the uptake and chemical resistance of the electrolyte. In-situ synthesis of inorganic nanoparticles in polymer solutions by sol–gel chemistry enables the nanoparticles to be well dispersed in the polymer and strongly bonded to it. This method is useful for incorporating silica, zirconia, or titania nanoparticles, which improve ionic absorption and chemical resistance.
3. **Electrospun Composites** Electrospinning provides a promising approach for the addition of nanoparticles or secondary polymers. Composite electrospun films present layered pore structures and good wettability in electrolytes. For example, the addition of ceramic nanoparticles in PAN/PVDF nanofibers provides films with good dimensional stability and ionic conductivity.

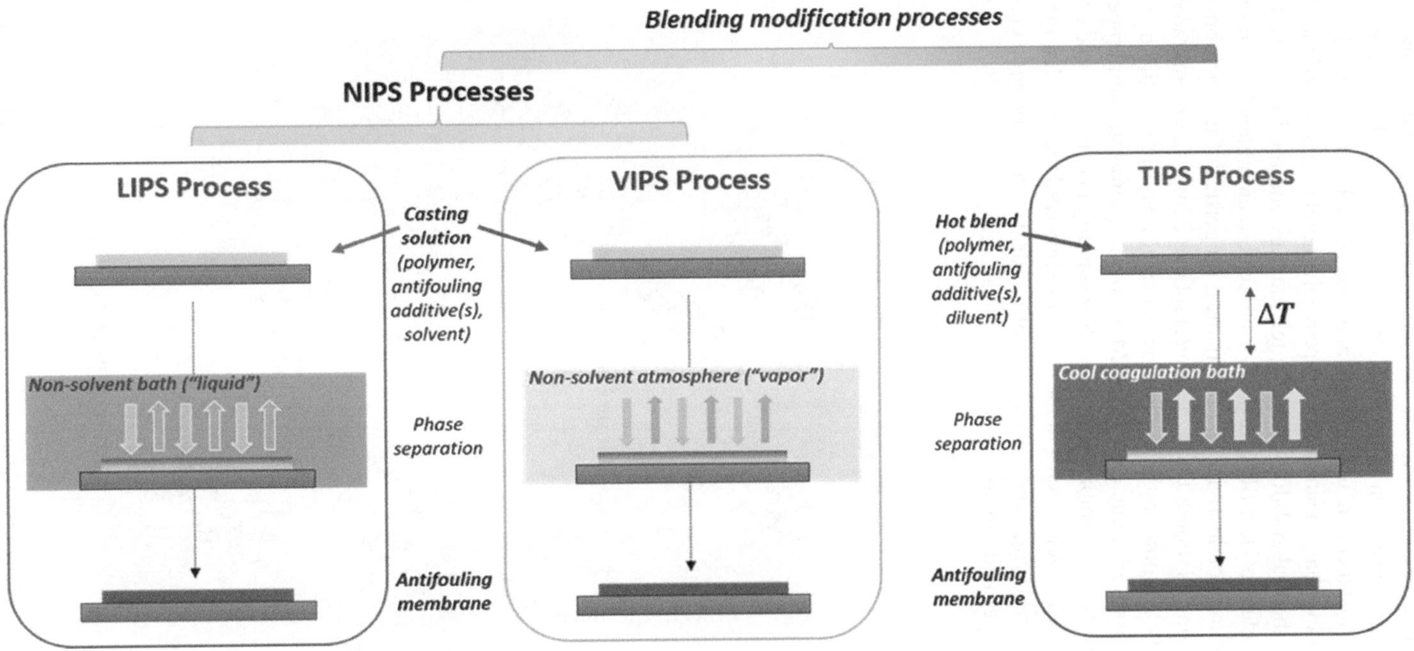


**FIGURE 3.5** Schematic representation of phase-inversion processes for the formation of flat-sheet antifouling porous membranes via blending modification.

**(Reproduced with permission [32].)**

The various types of membranes, including inorganic fillers, matrices of polymers, organic frameworks, MOFs, Covalent organic frameworks (COFs), and carbon-based nanomaterials, have been utilised to improve the electrochemical properties (Figure 3.6). For instance, in lithium-ion batteries, PVDF-$Al_2O_3$ and PAN-$SiO_2$ composite membranes have been used to improve the uptake capacity of the electrolyte, increase the thermal stability of the battery, and extend the life of the battery. In solid polymer electrolytes, SPEEK-$TiO_2$ composite membranes have been used to improve the proton conductivity and mechanical properties of the membranes for fuel cell applications. Nafion/$SiO_2$ or Nafion/graphene composite membranes have been used in RFB to improve the coulombic efficiency and life of the battery by minimising the crossover of vanadium ions. Metal oxide composite membranes have been used in electrolysers to improve the oxidative stability in extreme ionic environments [33]. The latest developments in the field have involved the introduction of 2D materials like MXenes, graphene oxide, and boron nitride, which provide enhanced mechanical properties and direction ion transport capacity.

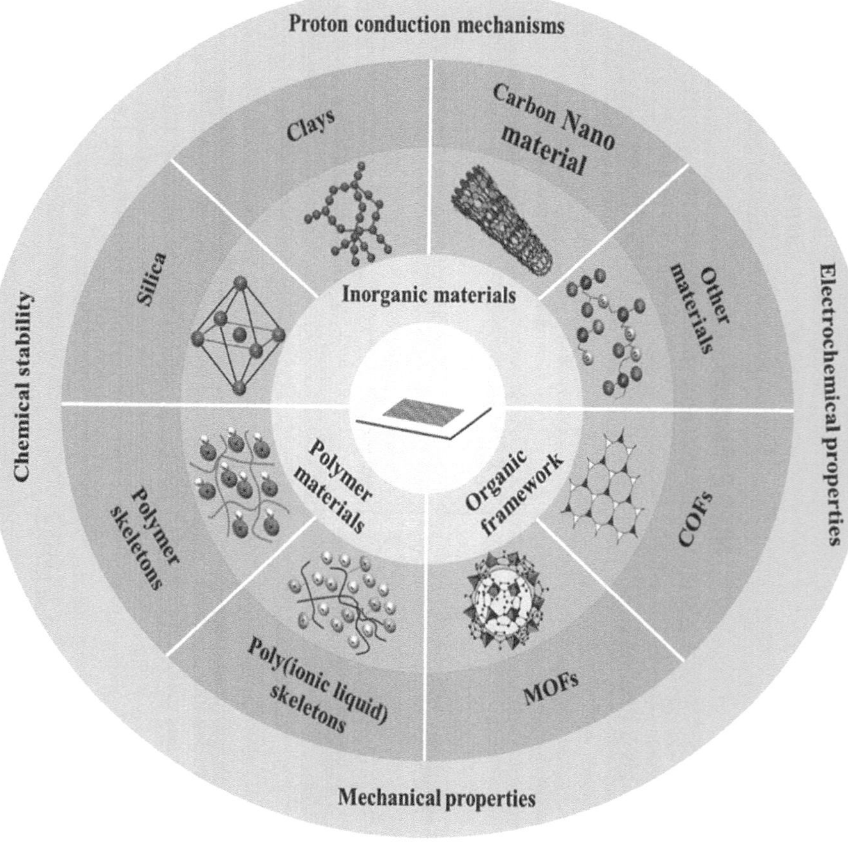


**FIGURE 3.6** Classification of membrane materials and their associated proton conduction, mechanical, electrochemical, and chemical stability characteristics.

New functional fillers, like sulfonated silica and phosphonic acid-functionalized $TiO_2$, have been designed as materials with the ability to act as ion conduction centres. The next possible approach is the utilisation of composite materials like PMOFs and taking advantage of the ion conduction and catalytic properties of MOFs with their unique frameworks. Hierarchical composite materials utilise different fillers, like ceramics, 2D materials, and polymers, for different purposes. Interface engineering utilises surface-functionalised fillers to reduce interfacial resistance and improve filler dispersion [34]. Green processing routes, such as aqueous solution-based sol–gel processing routes or even mechanical blending routes, are beneficial for a more sustainable approach. The addition of stimuli-responsive fillers, e.g., thermo-, pH-, or redox-responsive fillers, allows for the creation of smart membranes during the composite membrane fabrication process [35]. The composite approach of combining materials with different properties provides a promising route to create robust membranes with specific properties, which are needed to fulfil the stringent demands of the next-generation electrochemical devices. The characteristics of membrane fabrication methods for electrochemical energy devices are summarized in Table 3.1.

**TABLE 3.1**
**Comparison of Fabrication Methods for Membranes in Electrochemical Energy Devices**

| Fabrication Method | Principle | Advantages | Limitations | Applications in Electrochemical Devices |
|---|---|---|---|---|
| Solution Casting & Phase Inversion | Phase separation in polymer solutions | Simple, scalable, versatile | Limited nanoscale control, solvent issues | LIB membranes, PEMs, IEMs |
| Extrusion & stretching | Mechanical stretching of semicrystalline polymers | Industrial scalability, low cost | Poor wettability, thermal shrinkage | Polyolefin LIB membranes |
| Electrospinning | Electrohydrodynamic fibre formation | High porosity, tunable morphology | Weak mechanical strength, scale-up issues | LIBs, solid-state batteries, fuel cells |
| 3D Printing | LbL additive manufacturing | Precise control, hierarchical structures | Limited materials, slow speed | Batteries, electrolysers, flow batteries |
| LbL Assembly | Sequential deposition of layers | Nanoscale control, multifunctionality | Time-consuming, weak adhesion | Fuel cells, RFBs, LIBs |
| Composite/ Hybrid Fabrication | Blending polymers with fillers/ nanomaterials | Enhanced conductivity, stability | Dispersion challenges, cost | LIBs, fuel cells, electrolysers |

## 3.10 PROCESS–STRUCTURE–PROPERTY RELATIONSHIP

One of the most important aspects of membrane science is the interrelationship between the process of fabrication, the structure of the membrane, and the properties of the membrane. This implies that any process of fabrication, whether casting, stretching, electrospinning, or even 3D printing, imparts unique microstructural and chemical properties to the membranes, which are critical in defining the properties of the membranes in electrochemical systems. In this context, understanding the process, structure, and properties of membranes becomes critical in leveraging the properties of membranes in the rational design of membranes for specific applications in the energy sector. A variety of classes of membranes, including aromatic polymers, fluorinated polymers, modified polybenzimidazole membranes, and composite membranes, are some of the critical classes of membranes explored for the rational design of membranes with unique properties for advanced membrane fabrication (Figure 3.7).

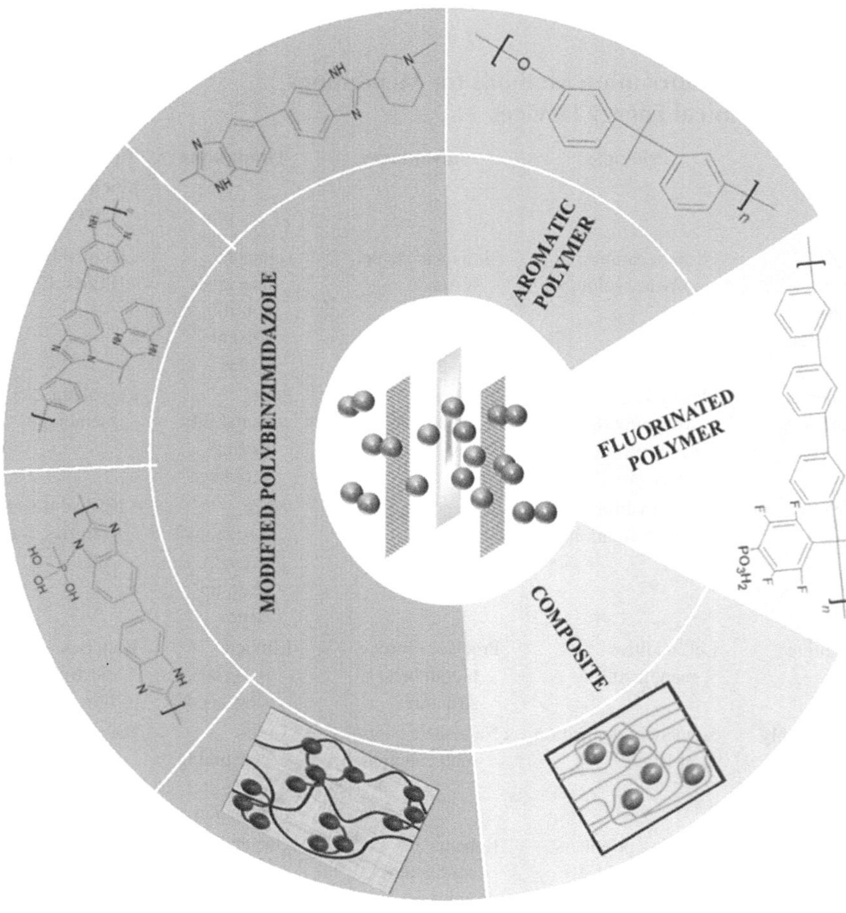


**FIGURE 3.7** Classification of polymer materials used in membrane fabrication.

### 3.10.1 Porosity and Pore Size Distribution

Porosity and pore size distribution are two very important structural characteristics that are highly dependent on the fabrication method. In the phase-inversion method, the pores are asymmetric and have a dense skin and a porous substructure. In the electrospinning method, the pores are highly porous and have a nanofibrous structure. This results in a high surface area and good uptake of the electrolyte. In the mechanical stretching method, the pores are slit-shaped, while in the track etching method, the pores are cylindrical and have a narrow size distribution. All the above characteristics are very important and highly dependent on the performance of the battery. High porosity results in good uptake and conductivity. However, the pores need to be optimised because high porosity results in poor mechanical strength. If the pores are too large, dendrite growth will occur in batteries. If the pores are too small, ionic resistance will occur.

### 3.10.2 Thickness and Dimensional Control

The thickness of the membranes has a crucial role to play in controlling ionic resistance, mechanical properties, and device performance. Fabrication methods such as solution casting and LbL assembly are used to precisely control the nanometre-scale thickness of the membranes, while extrusion and stretching methods are used to create thin membranes, which are essential for practical applications. In addition, new technologies such as 3D printing are also used to create membranes with specific thickness designs, depending on the application requirements. From a performance point of view, thin membranes are advantageous in terms of reduced ionic resistance and increased power density, while thick membranes are advantageous in terms of mechanical properties and safety, but at the expense of increased internal resistance, which negatively affects the efficiency of the device.

### 3.10.3 Mechanical Strength and Flexibility

Mechanical robustness, as well as flexibility, are crucial to ensuring membrane stability under various thermal, mechanical, and electrochemical stress conditions. Although electrospinning membranes are highly porous, they are usually supplemented with additional mechanical support owing to poor bonding between individual fibres. Composite fabrication methods, such as the addition of ceramic particles, are crucial in enhancing mechanical stiffness, thermal stability, and membrane integrity. Stretching methods are also effective in enhancing molecular alignment, thus increasing membrane mechanical strength as well as flexibility. Strong membranes are effective in withstanding thermal shrinkage, thus preventing short circuit formation, while flexible membranes are critical in wearable power devices, as well as solid electrolytes, which are subjected to mechanical deformation.

### 3.10.4 Functional Integration

Advanced fabrication techniques also facilitate the integration of multifunctional properties in membranes. Fabrication techniques of LbL are used for the deposition of ultrathin functional coatings, which may be ion-selective coatings, flame-retardant coatings, or catalytic coatings, for instance. Additionally, hybrid fabrication techniques, which may involve slot-die coating/electrospinning or sol–gel casting/casting, may be used for simultaneous control of bulk properties and functional coatings. These techniques facilitate the integration of various functionalities, which may include improved ion selectivity, flame retardancy, catalytic activity, and self-healing properties, for instance. Additionally, functional integration may be used to improve the long-term stability of membranes by eliminating degradation mechanisms such as gas crossover, dendrite growth, and radicals, for instance.

### 3.10.5 Chemical Compatibility

Chemical compatibility is essential in achieving membrane stability over extended periods in harsh operating environments. The membrane fabrication methods, such as phase inversion and sol-gel processing, enable the achievement of chemical stability through the incorporation of chemically stable inorganic fillers that can withstand harsh chemical environments such as oxidation, acid, and alkali. In the case of LbL films, the coatings can be tailored to introduce hydrophilic, proton-conducting, and oxidation-resistant surface properties. Incompatible materials can cause swelling, cracking, and dissolution, leading to membrane failure.

### 3.10.6 Multi-scale Architectures

Currently, modern hybrid technologies are capable of creating membranes with hierarchical pore size distributions, which may feature macro-, micro-, and nanoscale components. These membranes are advantageous in that it is possible to optimise ionic conductivity, electrolyte uptake, and strength in these membranes, which is generally not achievable in membranes of any of these scales individually. Table 3.2

**TABLE 3.2**
**Process–Structure–Property Relationship in Membrane Fabrication**

| Fabrication Method | Structural Features | Influencing Properties | Effect in Electrochemical Devices |
|---|---|---|---|
| Phase Inversion | Asymmetric pore structure | Ionic conductivity, selectivity | Improves electrolyte uptake in LIBs |
| Stretching / Track Etching | Fibrillar or cylindrical pores | Mechanical strength, thermal stability | Enhances dimensional stability |
| Electrospinning | Nanofiber network | Porosity, wettability | Enables fast ion transport in separators |
| 3D Printing | Hierarchical and patterned geometry | Thickness control, ion pathways | Enhances uniform ion flux and dendrite suppression |
| LbL Assembly | Nanoscale multilayer coatings | Ion selectivity, surface functionality | Reduces crossover in flow batteries |
| Composite Fabrication | Filler–polymer interfaces | Mechanical and thermal stability | Improves safety and cycling stability |

illustrates the process–structure–property relationships governing membrane performance in electrochemical energy devices.

## 3.11 EMERGING AND HYBRID TECHNIQUES

While traditional methods such as casting, stretching, and electrospinning are commonly used methods for the fabrication of membranes, the need for multifunctional and efficient membranes has led to the development of novel and hybrid fabrication methods. This has resulted in the combination of the characteristics of well-established fabrication methods and novel fabrication methods. This has resulted in the development of novel and efficient shapes and functions in the membranes. The novel and efficient shapes and functions are the result of the combination of the characteristics and the fabrication methods. This has resulted in the development of novel and efficient shapes and functions in the membranes. The novel and efficient shapes and functions are the result of the combination of the characteristics and the fabrication methods. This has resulted in the development of novel and efficient shapes and functions in the membranes. The novel and efficient shapes and functions are the result of the combination of the characteristics and the fabrication methods. Ink-Jet Method Surface Modification: The ink-jet method is used for the modification of the surface of the membrane by printing the desired coatings on the surface.

For example, PVDF membranes were functionalised with tannic acid or dopamine to attain significantly enhanced hydrophilicity, electrolyte uptake, and interfacial compatibility. This approach is easy to scale up and can be easily integrated into an existing roll-to-roll process line. Slot-Die Coating: This process is widely used in the battery and fuel cell industries for electrode functionalisation. It has recently been utilised for the functionalisation of membranes as well. It is utilised for the deposition of thin and uniform films on a large scale. The same process can be utilised for the deposition of ceramic-polymer composite films, flame-retardant films, and ion-conductive films on the membranes with high accuracy and precision. However, the high capital and operating costs of this process might restrict its application for scaling up the process for advanced membrane fabrication. Hybrid Phase Separation Methods: These methods combine TIPS and VIPS, or phase inversion and leaching, to create membranes with hierarchical porosity. These membranes are characterised by the presence of macro-, micro-, and nanopores in one membrane material. Such membrane materials are of great importance for the fabrication of SSEs and flow batteries, where ion and charge carrier transport is critical for the performance of the material.

## 3.12 APPLICATIONS IN ELECTROCHEMICAL SYSTEMS

Some of the applications in electrochemical systems involve the use of hybrid and advanced membranes to improve the performance of different devices. In the case of lithium-ion batteries, the application of hybrid membranes with ceramic coatings using slot-die and ink-jet methods improves the thermal stability and wettability of the devices. In the case of fuel cells, rotating spinneret 3D-printed membranes with channels improve the proton conductivity of the devices. In the case of redox flow batteries, the application of hybrid phase-inversion and ink-jet membranes improves the crossover of active species while maintaining the ionic conductivity of the devices. In the case of electrolysers, slot-die-coated membranes with embedded catalytic nanoparticles improve the efficiency and gas evolution of the devices. Some recent developments include the design of multifunctional coatings that can be used for the simultaneous achievement of flame resistance, ionic selectivity, and mechanical reinforcement in a single hybrid material layer. In electrolysers, slot-die-coated catalytic layers can be used to achieve gas evolution and durability in electrolysers [36]. Recent developments in this area have focused on the design of multifunctional coatings, which can be used to achieve the simultaneous realisation of three functionalities: namely, flame resistance, ionic selectivity, and mechanical reinforcement in a unified hybrid structure. In parallel, the use of 2D nanomaterials, such as graphene oxide, MXenes, and layered clays, in hybrid coatings has enabled the realisation of membranes with tunable ion channels and high electrochemical stability. In parallel, bio-inspired hybrid approaches, such as those inspired by ion channels or layered structures, are also gaining popularity for the realisation of membranes with selective and adaptive properties.

Such advances suggest the possibility of the development of fully integrated continuous manufacturing systems, which incorporate all of the above technologies, namely, electrospinning, slot-die coating, as well as surface modification, into a single processing platform. In addition, green materials, which are sustainable, could be considered as hybrid coating materials, which in turn could reduce the usage of harmful solvents as well as additives. Another possible area of discussion could be the potential of creating programmable adaptive hybrid membranes, which could react to external stimuli, such as temperature, pH, and electric fields, to create smart materials for energy devices. AI technology has the potential to function as a tool in the creation of hybrid fabrication parameters, which in turn could result in the fast discovery of new hybrid membranes. Emerging as well as hybrid technologies are at the centre of laboratory-scale innovations as well as industrial-scale production, thus offering a platform of great potential, namely, the next-generation membranes, which are not only stable, efficient, multifunctional, sustainable, but also adaptable to the varied requirements of the next-generation energy technologies.

## 3.13 CONCLUSION

The fabrication of membranes for electrochemical energy systems has progressed from conventional techniques such as solution casting, phase inversion, and extrusion to advanced techniques including electrospinning, 3D printing, LbL assembly, and hybrid techniques. Although conventional techniques remain prominent in membrane fabrication for electrochemical systems due to their reliability and lower production costs, advanced techniques improve ionic conductivity, mechanical properties, chemical stability, and overall electrochemical device performance. The introduction of new materials such as 2D nanostructures, ionic liquids, and bio-inspired polymers, along with innovative fabrication practices, is paving the way for a new era in membrane materials for electrochemical systems, where membranes can be used for enhancing efficiency, safety, and longevity in these systems.

## REFERENCES

[1] J. Ye, L. Xia, H. Li, F. P. G. de Arquer, and H. Wang, "The critical analysis of membranes toward sustainable and efficient vanadium redox flow batteries," *Adv. Mater.*, vol. 36, no. 28, pp. 1–24, 2024, doi: 10.1002/adma.202402090

[2] M. D. Bouguern, A. K. Madikere Raghunagtha Reddy, and K. Zaghib, "The critical role of interfaces in advanced Li-ion battery technology: A comprehensive review," *J. Power Sources*, vol. 623, no. July, p. 235457, 2024, doi: 10.1016/j.jpowsour.2024.235457

[3] N. Khan, C. A. Ooi, A. Alturki, M. Amir, and T. Alharbi, "A critical review of battery cell balancing techniques, optimal design, converter topologies, and performance evaluation for optimizing storage system in electric vehicles," *Energy Reports*, vol. 11, no. April, pp. 4999–5032, 2024, doi: 10.1016/j.egyr.2024.04.041

[4] R. Abdulhussain, A. Adebisi, B. R. Conway, and K. Asare-Addo, "Electrospun nanofibres: Exploring process parameters, polymer selection, and recent applications in pharmaceuticals and drug delivery," *J. Drug Deliv. Sci. Technol.*, vol. 90, no. August, p. 105156, 2023, doi: 10.1016/j.jddst.2023.105156

[5] A. Nowik-Zając and V. Sabadash, "Recent developments in polymer inclusion membranes: Advances in selectivity, structural integrity, environmental applications and sustainable fabrication," *Membranes*, vol. 15, no. 8, p. 249, 2025, doi: 10.3390/membranes15080249

[6] P. D. Dalton, C. Vaquette, B. L. Farrugia, T. R. Dargaville, T. D. Brown, and D. W. Hutmacher, "Electrospinning and additive manufacturing: Converging technologies," *Biomater. Sci.*, vol. 1, no. 2, pp. 171–185, 2013, doi: 10.1039/c2bm00039c

[7] J. Lipton, G. M. Weng, J. A. Röhr, H. Wang, and A. D. Taylor, "Layer-by-layer assembly of two-dimensional materials: Meticulous control on the nanoscale," *Matter*, vol. 2, no. 5, pp. 1148–1165, 2020, doi: 10.1016/j.matt.2020.03.012

[8] S. Mazinani, S. Darvishmanesh, A. Ehsanzadeh, and B. Van der Bruggen, "Phase separation analysis of Extem/solvent/non-solvent systems and relation with membrane morphology," *J. Memb. Sci.*, vol. 526, no. December 2016, pp. 301–314, 2017, doi: 10.1016/j.memsci.2016.12.031

[9] S. Harinivalli, S. Aparna, A. K. A. S., K. Arockiaveronikka, P. Kaleeswari, N. Raja, G. Arthanareeswaran, and R. V. Mangalaraja, Hybrid adsorption–nanofiltration using modified pei-based mixed matrix membranes for efficient dye rejection from wastewater. *Water, Air, Soil Pollution*, vol. 236, no. 13, pp. 1–23, Dec. 2025, doi: 10.1089/ten.teb.2009.0639

[10] A. Kausar, "Phase inversion technique-based polyamide films and their applications: A comprehensive review," *Polym. - Plast. Technol. Eng.*, vol. 56, no. 13, pp. 1421–1437, 2017, doi: 10.1080/03602559.2016.1276593

[11] S. Favero, I. E. L. Stephens, and M. Titirci, "Anion exchange ionomers: Design considerations and recent advances - an electrochemical perspective," *Advanced Materials*, vol. 36, no. 8, Feb. 2024, doi: 10.1002/adma.202308238

[12] K. Staszczyk and R. Tylingo, "Bio-based proton exchange membranes from chitosan: A review of progress and challenges," *Energy and Fuels*, 2025, doi: 10.1021/acs.energyfuels.5c01969

[13] A. Bin Rashid, M. Haque, S. M. M. Islam, and K. M. R. Uddin Labib, "Nanotechnology-enhanced fibre-reinforced polymer composites: Recent advancements on processing techniques and applications," *Heliyon*, vol. 10, no. 2, p. e24692, 2024, doi: 10.1016/j.heliyon.2024.e24692

[14] L. Suamte, A. Tirkey, J. Barman, and P. J. Babu, "Various manufacturing methods and ideal properties of scaffolds for tissue engineering applications," *Smart Mater. Manuf.*, vol. 1, no. 1, p. 100011, Jan. 2023, doi: 10.1016/j.smmf.2022.100011

[15] R. Huang, X. Zhu, H. Tu, and A. Wan, "The crystallization behavior of porous poly(lactic acid) prepared by modified solvent casting/particulate leaching technique for potential use of tissue engineering scaffold," *Mater. Lett.*, vol. 136, pp. 126–129, 2014, doi: 10.1016/j.matlet.2014.08.044

[16] S. Murugan and S. R. Parcha, "Fabrication techniques involved in developing the composite scaffolds PCL/HA nanoparticles for bone tissue engineering applications," *J. Mater. Sci. Mater. Med.*, vol. 32, no. 8, pp. 1–10, 2021, doi: 10.1007/s10856-021-06564-0

[17] C. Sanchez, B. Julián, P. Belleville, and M. Popall, "Applications of hybrid organic-inorganic nanocomposites," *J. Mater. Chem.*, vol. 15, no. 35–36, pp. 3559–3592, 2005, doi: 10.1039/b509097k

[18] A. Popova, T. Shintani, and T. Fujioka, "Track-etched membrane as a thin substrate with straight pores to fabricate polyamide forward osmosis membrane," *J. Membr. Sci. Lett.*, vol. 4, no. 1, p. 100068, Jun. 2024 1, doi: 10.1021/acs.chemrev.8b00593

[19] K. Bicy, N. Kalarikkal, A. M. Stephen, D. Rouxel, and S. Thomas, "Facile fabrication of microporous polypropylene membrane separator for lithium-ion batteries," *Mater. Chem. Phys.*, vol. 255, no. March, p. 123473, 2020, doi: 10.1016/j.matchemphys.2020.123473

[20] S. C. Mun and J. H. Won, "Manufacturing processes of microporous polyolefin separators for lithium-ion batteries and correlations between mechanical and physical properties," *Crystals*, vol. 11, no. 9, pp. 1–14, 2021, doi: 10.3390/cryst11091013

[21] A. A. Mashentseva, D. S. Sutekin, S. R. Rakisheva, and M. Barsbay, "Composite Track-Etched Membranes: Synthesis and Multifaced Applications," *Polymers*, vol. 16, no. 18, p. 2616, 2024, doi: 10.3390/polym16182616

[22] T. Ma, J. M. Janot, and S. Balme, "Track-etched nanopore/membrane: From fundamental to applications," *Small Methods*, vol. 4, no. 9, pp. 1–36, 2020, doi: 10.1002/smtd.202000366

[23] P. S. Kim, A. Le Mong, and D. Kim, "Thermal, mechanical, and electrochemical stability enhancement of Al2O3 coated polypropylene/polyethylene/polypropylene separator via poly(vinylidene fluoride)-poly(ethoxylated pentaerythritol tetraacrylate) semi-interpenetrating network binder," *J. Memb. Sci.*, vol. 612, no. February, p. 118481, 2020, doi: 10.1016/j.memsci.2020.118481

[24] K. Morikawa, A. Vashisth, C. J. Grimme, M. J. Green, and M. Naraghi, "Wire melt electrospinning of thin polymeric fibres via strong electrostatic field gradients," *Macromol. Mater. Eng.*, vol. 304, no. 1, pp. 1–9, 2019, doi: 10.1002/mame.201800417

[25] M. Ahmadi Bonakdar and D. Rodrigue, "Electrospinning: Processes, structures, and materials," *Macromol*, vol. 4, no. 1, pp. 58–103, 2024, doi: 10.3390/macromol4010004

[26] M. Cai et al., "Lithium ion battery separator with improved performance via side-by-side bicomponent electrospinning of PVDF-HFP/PI followed by 3D thermal crosslinking," *J. Power Sources*, vol. 461, no. September 2019, pp. 1–10, 2020, doi: 10.1016/j.jpowsour.2020.228123

[27] V. Mottaghitalab, G. M. Spinks, and G. G. Wallace, "The influence of carbon nanotubes on mechanical and electrical properties of polyaniline fibres," *Synth. Met.*, vol. 152, no. 1–3, pp. 77–80, 2005, doi: 10.1016/j.synthmet.2005.07.154

[28] J. Zhao et al., "Recent advances in coaxial electrospun nanofibres for wound healing," *Mater. Today Bio*, vol. 29, no. September, p. 101309, 2024, doi: 10.1016/j.mtbio.2024.101309

[29] S. Mametja, M. Mabuza, E. R. Sadiku, R. S. Mohlamonyane, and J. S. Sefadi, *Development of Energy-efficient Polymers by Using Conductive Nanohybrid Fillers: Recent Progress and Future Prospects*, vol. 8, no. 3. Springer International Publishing, 2025. doi: 10.1007/s42114-025-01330-0

[30] Y. X. Yan, H. B. Yao, L. B. Mao, A. M. Asiri, K. A. Alamry, H. M. Marwani, and S. H. Yu, "Micrometer-thick graphene oxide–layered double hydroxide nacre-inspired coatings and their properties," *Small.*, vol. 12, no. 6, pp. 745–755, 2016 Feb. doi: https://doi.org/10.1002/smll.201502061
[31] S. Zhang, F. Xia, S. Demoustier-Champagne, and A. M. Jonas, "Layer-by-layer assembly in nanochannels: Assembly mechanism and applications," *Nanoscale*, vol. 13, no. 16, pp. 7471–7497, 2021, doi: 10.1039/d1nr01113h
[32] T. A. Geleta, I. V. Maggay, Y. Chang, and A. Venault, "Recent advances on the fabrication of antifouling phase-inversion membranes by physical blending modification method," *Membranes (Basel).*, vol. 13, no. 1, 2023, doi: 10.3390/membranes13010058
[33] J. Balaji, M. G. Sethuraman, S. H. Roh, and H. Y. Jung, "Recent developments in sol-gel based polymer electrolyte membranes for vanadium redox flow batteries – A review," *Polym. Test.*, vol. 89, no. May, 2020, doi: 10.1016/j.polymertesting.2020.106567
[34] F. Z. Zeggai, Z. Ait-Touchente, K. Bachari, and A. Elaissari, "Investigation of metal-organic frameworks (MOFs): Synthesis, properties, and applications - an in-depth review," *Chem. Phys. Impact*, vol. 10, no. March, 2025, doi: 10.1016/j.chphi.2025.100864
[35] M. Bril, S. Fredrich, and N. A. Kurniawan, "Stimuli-responsive materials: A smart way to study dynamic cell responses," *Smart Mater. Med.*, vol. 3, no. December 2021, pp. 257–273, 2022, doi: 10.1016/j.smaim.2022.01.010
[36] C. Hu et al., "Advanced patterned membranes for efficient alkaline membrane electrolyzers," *ACS Energy Lett.*, vol. 9, no. 3, pp. 1219–1227, 2024, doi: 10.1021/acsenergylett.4c00207

# 4 Properties of Membranes in Electrochemical Devices

## 4.1 INTRODUCTION

The operation of electrochemical cells is based on the synergistic integration of physical, chemical, mechanical, ionic, thermal, and hygroscopic properties [1]. All these properties play an important role in the operation of electrochemical cells. For example, physical properties such as thickness, porosity, pore size, and shape affect ionic conductivity, electrolyte uptake, and transport properties [2]. Additionally, chemical properties like composition, degree of crystallinity, polarity, and compatibility with electrolytes play a role in determining the electrochemical stability of the membrane as well as the reactivity of the polymer in the electrochemical environment. Mechanical properties such as tensile strength, elongation at break, and puncture strength enable the membrane to withstand its structural integrity without shrinking, breaking, or short-circuiting. Ionic properties such as ionic conductivity, transference number of cations/anions, and selectivity are significant in the efficient transport of ions and minimizing unnecessary ion transport [3]. Thermal properties such as melting point, thermal shrinkage resistance, and flame retardancy are also important for safety and durability when the material is subjected to extreme operating conditions. For proton-exchange membrane fuel cells and proton-exchange membrane flow batteries, the material's hygroscopic properties, such as water absorption and swelling, are particularly relevant to the material's conductivity and lifetime. The overall membrane properties for LIBs, fuel cells, electrolyzers, RFBs, etc., must be balanced effectively. Quantification of these properties via conventional techniques such as tensile tests, TGA, impedance spectroscopy, and water uptake measurements can provide a deeper understanding of membrane behaviour when operating in real-world scenarios. In this overview, the main properties of membranes for electrochemical devices will be addressed, including their definitions, characterization techniques, and influence on the overall performance of the devices. It is the hope that a balance between material properties and real-world performance will provide a set of guidelines for the rational design of membranes that can satisfy the requirements for the next generation of electrochemical devices.


DOI: 10.1201/9781003679363-4

## 4.2 PHYSICAL PROPERTIES

### 4.2.1 Thickness

The Thickness of the Membrane is an important parameter that has a direct impact on ionic resistance, energy efficiency, and mechanical stability. According to Ohm's law, the ionic resistance is directly proportional to the thickness. Therefore, if the membrane is thinner, there would be lower ohmic losses. As a result, there would be an increase in the power density of electrochemical devices so that they are able to attain a higher power density with lower polarization losses [4]. The thin membranes measuring between 10 and 20 μm are also being taken into consideration, and it is necessary to reduce the ionic resistance, which is vital for their optimal performance, especially in the case of proton-exchange membrane fuel cells (PEMFCs) and electrodialysis. For example, Nafion membranes with thicknesses of 10–20 μm exhibit reduced ionic resistance and enhanced proton conductivity under adequately hydrated conditions, thereby improving the power density and efficiency of portable and automotive fuel cells. In solid-state electrolytes, thinner membranes exhibit lower contact resistance between the electrode and the electrolyte. However, certain trade-offs have to be made for the use of thinner membranes. For instance, ultra-thin membranes have a higher chance of mechanical failure, pinholes, dimensional instability, and dendrite penetration, especially for high-voltage lithium-ion batteries and redox flow batteries. Therefore, the membranes have to be thick to withstand hydrostatic pressures, swelling forces, and chemical effects after long periods of use. Thicker membranes with a thickness of 50–200 μm have also been used for flow batteries to prevent crossover effects [5]. In contrast, lithium-ion battery membranes are usually 20–30 μm thick to achieve optical ionic conductivity, mechanical strength, and safety. Recently, innovations such as thin membranes reinforced with nanofibrous mats, woven fabrics, or ceramic fillers have emerged. These offer mechanical support and, at the same time, ensure that the functional layer is as thin as possible [6]. In addition, multilayer membranes are designed with a strong porous scaffold that supports an ultra-thin ion-conductive layer, thus decoupling membrane thickness from mechanical strength. Finally, the optimal membrane thickness will depend on the device, and there will be a trade-off among ionic conductivity, selectivity, mechanical integrity, and chemical stability. The future of electrochemical systems lies in the production of thinner, more conductive, and durable membranes. This can be achieved by composite materials, nanostructured materials, and hybrid manufacturing techniques.

### 4.2.2 Porosity and Pore Size

Porosity and pore size distribution are among the most important structural parameters of electrochemical devices. These parameters have a great effect on the efficiency of electrochemical devices. The parameters include absorption of the electrolyte, permeability, selectivity, and dimensional stability of the devices. Porosity is an indicative parameter of the percentage of voids contained within the membrane. High porosity values have generally been known to have a reducing effect on ionic resistance, hence enhancing ionic conductivity and power density [7]. For lithium-ion battery separators, a porosity of approximately 30–60% is generally considered optimal, whereas most commercial separators possess porosity values between 40% and 50%, balancing electrolyte wettability, ionic conductivity, and mechanical strength. The trade-off for the mechanical properties is quite high.

In the case of flow batteries and fuel cells, the presence of large or uncontrolled pores in the membranes allows the active species to cross over from one electrode compartment to the other [8]. This leads to chemical reactions that are not beneficial for the generation of electricity, thereby reducing the coulombic efficiency and increasing the rate of undesirable chemical reactions such as degradation. The pore size and pore shape are the fundamental characteristics that are of equal significance. The pore size distribution model and the aspect ratio are essential for the optimization of permeability and selectivity. The small pore size ensures that larger solvated ions or redox species cannot cross over, yet allows for fast ionic conductivity [9]. On the other hand, larger or elongated pores, which are commonly observed in stretched polyolefin membranes, usually offer higher permeability but lower selectivity.

The pore structure is very sensitive to the membrane fabrication method. Phase inversion allows the formation of asymmetric membranes with a distribution of pore size, while electrospinning produces highly interconnected nanofiber networks with adjustable porosity. Deformation of semicrystalline polymers such as PE or PP causes slit-like pores to form at the interface between crystalline and amorphous regions. Track-etching allows the formation of nearly monodisperse cylindrical pores with precisely controlled diameters, thereby offering high selectivity.

However, recent advances in porosity have made it feasible to design hierarchically structured pore networks that can synergistically integrate macro-, meso-, and micropores to enhance ionic conductivity, mechanical reinforcement, and transport selectivity in a single system. For example, nanofiber membranes with precisely controlled microporous coatings can provide efficient ion transport and prevent the penetration of dendrites in lithium-metal batteries. Similarly, there has been a recent trend in the development of hybrid membranes that incorporate inorganic fillers such as $SiO_2$, $Al_2O_3$, or clays in an effort to enhance the stability of the pore structures with respect to heat or stress. The ability to precisely control these parameters through fabrication methods, additive use, and hybrid designs is fundamental to optimizing membranes for various electrochemical systems, including lithium-ion batteries, flow batteries, and fuel cells [10].

### 4.2.3 Morphology and Microstructure

Membrane morphology and microstructure are important parameters that have a significant influence on transport properties, interfacial configuration, and stability of an electrochemical device [11]. Unlike membrane thickness and porosity, which are scalar parameters, membrane morphology is a description of the spatial arrangement of polymer chains and domains, fillers, and surface and bulk structural features [12]. Normally, techniques like scanning SEM, AFM, and TEM are used to characterize membrane morphology at various scales. They help in understanding various parameters like pore geometry, inter-connectivity, phase separation and fibre orientation. All these parameters are correlated with electrochemical performance. Homopolymer membrane morphology is normally controlled by crystallinity, chain packing, and phase separation during the process of solidification [13]. High performance is normally associated with the presence of well-defined microphase-separated channels that aid in selective ion transport. Furthermore, it is also important to control the dispersion of inorganic fillers in polymer-ceramic nanocomposites, like those of $SiO_2$, $TiO_2$, $Al_2O_3$, MOFs, or clays, in order to avoid their aggregation, as this might cause non-homogeneous conductivity, mechanical problems, or pore closing [14]. The homogeneous distribution of inorganic fillers in the matrix is useful in optimizing the dimensional stability, reducing thermal shrinkage, and enhancing the chemical resistance of the material. Block copolymer membranes are one of the examples of the wide range of applications of morphology control, in which microphase separation channels are sought in order to optimize ionic conductivity, despite difficulties in microphase separation. Moreover, the surface morphologies of electrospun nanofibrous membranes, with high surface-to-volume ratios, aligned or random fibre orientations, and porous networks, enable fast penetration of electrolytes. Surface roughness or texture can enhance electrolyte wetting properties or adhesion to electrodes, as determined by AFM studies, whereas dense films, such as layer-by-layer films, can offer selectivity or resistance [15]. Hybrid membranes, in which there is a combination of porous structure and surface functionalities, are also considered in LIBs, VRFBs, etc [16]. The level of progress that has been made in the development of in situ imaging techniques, such as synchrotron X-ray tomography, cryo-TEM, and computational modelling, is a significant factor in understanding microstructural changes linked to long-term performance degradation. By employing optimal microstructure designs and advanced characterization methods, it is possible to develop membranes that offer the highest possible performance and stability for a wide range of electrochemical devices. Table 4.1 highlights the important properties of a membrane and the roles in performance and application of various electrochemical devices.

**TABLE 4.1**
**Summary of Key Membrane Properties and Their Impact on Device Performance**

| Property | Measurement Techniques | Impact on Device Performance | Typical Range | Electrochemical Devices |
|---|---|---|---|---|
| Thickness | Micrometre, SEM | Affects ionic resistance and durability | 10–200 μm | PEM fuel cells, Water electrolyzers, Lithium-ion batteries, Redox flow batteries |
| Porosity | Mercury porosimetry, BET | Controls electrolyte uptake, selectivity | 30–60% (LIBs) | Lithium-ion batteries, Supercapacitors, Redox flow batteries |
| Ionic Conductivity | EIS | Determines efficiency, reduces ohmic loss | $10^{-4}$–$10^{-1}$ | Fuel cells, Solid-state batteries, Electrolyzers |
| Mechanical Strength | Tensile test | Ensures stability and safety | 20–200 MPa | Lithium-ion batteries, Sodium-ion batteries, Flow batteries |
| Thermal Stability | TGA, DSC | Prevents shrinkage, melting | >100°C (PEMFCs) | Fuel cells, Electrolyzers, High-energy LIBs |
| Water Uptake | Gravimetric | Affects swelling, proton transport | 10–30% | Fuel cells, Electrolyzers, Vanadium redox flow batteries |

## 4.3 IONIC CONDUCTIVITY AND SELECTIVITY

Ionic conductivity is one of the most important factors that indicate a membrane's performance in an electrochemical device because it primarily determines the overall efficiency of the cell, as illustrated in Figure 4.1. It is usually expressed in siemens per centimetre (S/cm) and is dependent on a variety of parameters, including the chemistry of the polymer backbone, ion-exchange group density, degree of hydration (for proton membranes), porosity, and temperature. For PEMs like those used in fuel cells, a conductivity of at least $10^{-1}$ S/cm under operating conditions is generally regarded as the minimum for effective performance (Table 4.2). The fully hydrated membranes, such as Nafion and other PFSA-type membranes, reach such values because of their characteristic phase-segregated structure, in which hydrophilic ionic aggregates form conducting paths [18]. However, the conductivity decreases when the membranes are subjected to heat and low relative humidity. This has led researchers to look for new high-temperature PEMs, such as PA-doped PBI and sulfonated aromatic polymers. The ionic conductivity of SPEs employed in lithium-ion and solid-state batteries ranges from $10^{-4}$ S/cm to $10^{-2}$ S/cm at room temperature. These values are two orders of magnitude lower than the ionic conductivity of PEMs. High ionic conductivity in SPEs is difficult to attain due to the fact that the main mechanism of ion transport in SPEs is the segmental motion of the polymer chains, which is greatly dependent upon the glass transition temperature of the polymers and their degree of crystallinity. For instance, PEO-based electrolytes rely on the flexibility of the polymer chains to dissolve and move $Li^+$ ions, but at room temperature, high crystallinity keeps the conductivity relatively low [19].

These fillers enhance ionic conductivity through multiple mechanisms. They also reduce polymer crystallinity, allowing for an increase in ion mobility, and the fillers facilitate ion dissociation through Lewis acid-base interaction. Additionally, there are polymer filler interfaces that provide continuous pathways for fast ion conduction. The mechanical strength enhancement of the composite membrane is also beneficial in suppressing lithium dendrite growth. Temperature is also a critical factor: in most cases, ionic conductivity follows an Arrhenius behaviour, with conductivity increasing as temperature rises due to enhanced chain mobility and ion hopping. The design of high-temperature polymer electrolytes has been inspired by this behaviour; these electrolytes are stable at temperatures above 100°C and still maintain sufficient conductivity for applications such as all-solid-state lithium-metal batteries and high-temperature PEM fuel cells. Recently, studies have been exploring the interaction between polymeric ion conductors and ceramic frameworks (e.g., PEO-lithium lanthanum zirconium oxide (LLZO) composites, and SPEEK–MOF hybrids) to achieve high conductivity, mechanical reinforcement, and chemical stability, especially considering the potential degradation of the materials. These designs are among the closest attempts to bridge the gap between liquid electrolytes, which have very high ionic conductivity but are unsafe, and solid electrolytes, which are safer but less conductive. Ionic conductivity involves a trade-off: membranes need to have sufficiently high conductivity to minimize ohmic losses,

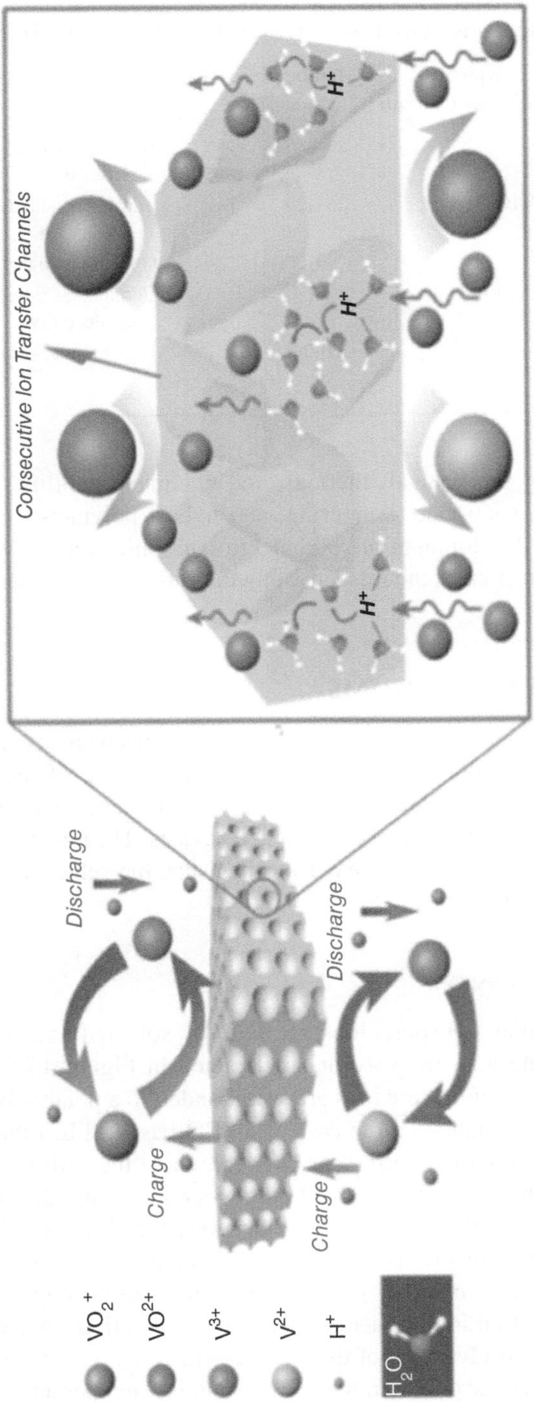


**FIGURE 4.1** Schematic illustration of ion transfer channels.

**(Reproduced with permission [17].)**

**TABLE 4.2**
**Comparison of Ionic Conductivities in Different Membrane Types**

| Membrane Type | Typical Conductivity (S/cm) | Temperature (°C) | Key Consideration |
|---|---|---|---|
| Nafion (PEM) | $10^{-1}$ | 80 | High proton conductivity; expensive |
| SPEEK | $10^{-2}$–$10^{-1}$ | 25–120 | Tunable conductivity via sulfonation |
| PBI (acid-doped) | $10^{-2}$ | 150–200 | Operates at high T, low humidity |
| PEO–$LiClO_4$ | $10^{-4}$–$10^{-3}$ | 25 | Common SPE for LIBs; low RT conductivity |
| LLZO-based solid electrolyte | $10^{-3}$–$10^{-2}$ | RT–300 | High thermal stability |

while also providing mechanical, thermal, and chemical stability. Advances in composite materials, nanoscale structuring, and hybrid polymer-ceramic designs have accelerated the development of solid electrolyte ionic conductivities towards levels approaching practical standards, enabling their use in next-generation energy systems.

## 4.4 TRANSPORT MECHANISMS

Ion transport in the membrane is governed by basic mechanisms that influence conductivity, selectivity, and the electrochemical performance of the membranes. Although these ideas are complex, it is essential to grasp the ideas for the design of membranes that are best for use in energy storage devices. The ion transport in these membranes takes place through two major mechanisms: the vehicular transport and the Grotthuss mechanism.

### 4.4.1 Vehicular Transport

In the case of vehicular transport, ions are initially solvated and move directly through the membrane with the solvent, as depicted in Figure 4.2. This type of transport occurs in a system where ions are well bonded to a polar solvent or functional group [21]. For instance, in the case of a PEO-based SPE, lithium ions can be transported by coordination with ether oxygen, where the entire solvated complex is moved with the conformation of the polymer segments. Similarly, in ion-exchange membranes, in the presence of water, ions like $Na^+$ or $K^+$ move with water molecules, and conductivity is largely dependent on the quantity of water present as well as the ions' mobility [22]. Vehicular transport is therefore very sensitive to segmental motion in polymers, temperature, as well as uptake of solvents and crystallinity. The disadvantage of this mechanism is that it is dependent on the movement of the solvent or polymer, meaning that at low temperatures, conductivity is lowered.

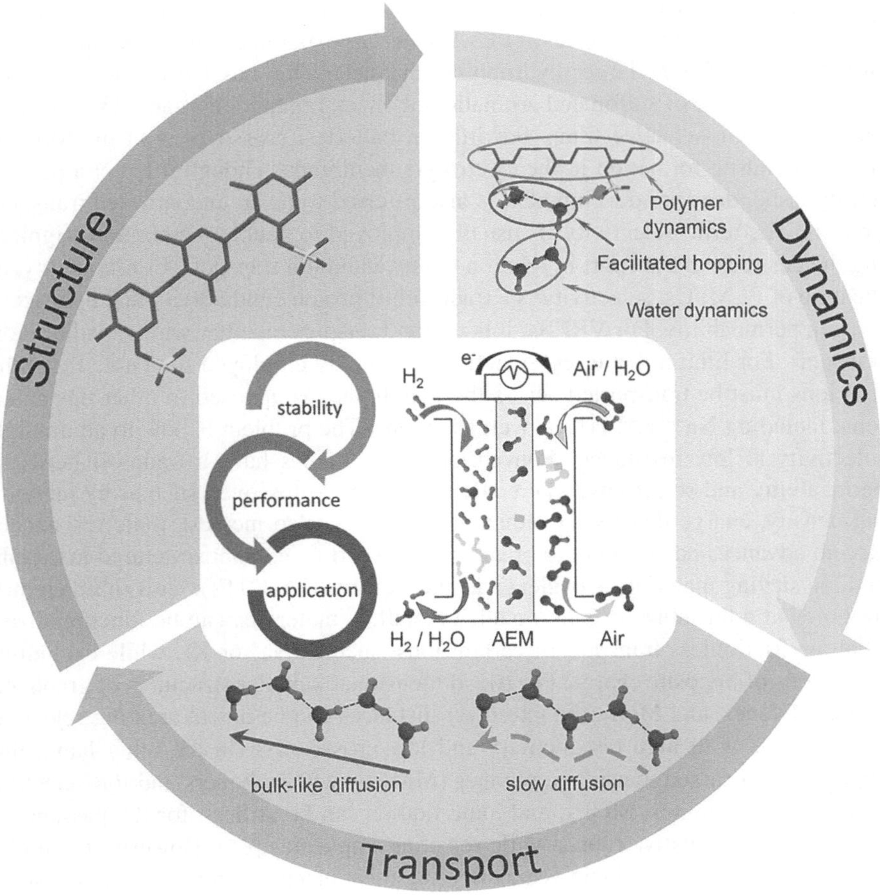


**FIGURE 4.2** Schematic representation of the interrelationship between structure, dynamics, and transport in ion-exchange membranes.

**(Reproduced with permission [20].)**

### 4.4.2 Permeability, Selectivity, and Crossover

The successful transport of ions through the membrane, while preventing the transport of other materials, is the basic necessity that determines the major features, efficiency, and longevity of the device. Parameters such as gas permeability, ionic selectivity, and crossover resistance have a direct effect on coulombic efficiency, fuel usage, and chemical stability. In PEMFCs, the minimization of hydrogen and oxygen crossover is a major design necessity [23]. Besides fuel efficiency loss, there is also the generation of reactive oxygen species, which cause damage to the polymer backbone. Nafion membranes, though having high proton conductivity, show some degree of gas permeability, especially when membranes are of low thickness. For VRFB, membranes must be effective in preventing vanadium ion transfer between the positive and negative electrolyte regions. Crossover results in electrolyte imbalance, capacity reduction, and decreased energy efficiency. Conventional membranes made of PFSA materials experience high vanadium crossover because of their larger hydrophilic channels. This has led to the design of membranes made of sulfonated aromatic polymers, organic–inorganic hybrids, and nanoporous ion-selective films. For lithium batteries, crossover is in the form of lithium dendrite formation in the electrolyte membrane. Though it is not a permeability issue, dendrite development is also referred to as an uncontrolled transport process. The term 'selectivity' is usually employed to denote membranes, signifying the desired ion transport in relation to unwanted ion transport. Consequently, in the case of PEMFCs, selectivity is a trade-off of proton conductivity and hydrogen/oxygen permeability. For VRFBs, it is proton transport together with vanadium ion transport. For lithium batteries, selectivity is usually employed to denote that lithium ions must be transported across the membrane, as opposed to other unwanted ions, including $Na^+$, $K^+$, $H^+$, as well as anions. The problem is how to attain high selectivity at low resistance. Conventional membranes have a trade-off between permeability and selectivity, where attempts to limit crossover, such as by increasing density, charge density, or reducing pore sizes, also increase ionic resistance. Recent advances have sought to break this trade-off using nanostructured materials and ion sieving materials. Covalent triazine frameworks (CTFs), with sub-nanometre pores of adjustable sizes in organic crystalline materials, can be functionalized to allow frictionless transport of certain ions, such as $Na^+$ or $K^+$, while excluding larger ions of opposite charge. For two-dimensional lamellar structures of graphene oxide, MXenes, and $MoS_2$, the interlayer distances can be tuned, creating selective ion channels with high conductivity and low crossovers. On the other hand, the application of mixed matrix membranes (MMMs) with polymers and ion-selective materials like zeolites, MOFs, and ionic liquids can be utilized for the passage of ions through alternative routes while rejecting impurities [24]. However, the problem of crossovers is the main issue affecting most electrochemical devices despite the recent breakthroughs. For PEMFCs, the crossovers enhance the rate of radical attacks by the membrane, while for VRFBs, crossovers result in reduced efficiency and the associated rebalancing costs [25]. In addition, in the case of lithium and zinc-based batteries, the unregulated transport of species is also responsible for the formation of dendrites and the risk of short circuits. Future membranes will need to balance structural resolution with functional selectivity to address the issue of permeability. Techniques such as nanoconfined channels, functionalized frameworks, and multilayer barriers are currently being considered as vital to the design of highly conductive, selective, and durable membranes.

### 4.4.3 Grotthuss Mechanism

The Grotthuss mechanism, also referred to as proton hopping, is a special type of transport mechanism whereby a series of steps of hydrogen bonding of a proton to a new site occur. In general terms, a proton can be transferred from one hydrogen bonding site to another [26]. This phenomenon can be better understood in a membrane such as a PEM, Nafion, or a sulfonated aromatic membrane, where a group of sulfonic acids acts as an acidic group for proton dissociation [27]. In hydrated membranes, water molecules form a cluster structure connected by hydrogen bonds. This allows for the rapid hopping of protons. The Grotthuss mechanism differs from the vehicular mechanism in its ability to allow for very high proton mobility, particularly under humid conditions. This mechanism can be used to explain the high conductivity of modern PEMs, which is $\geq 10^{-1}$ S/cm [27]. For instance, vehicular transport allows for cation conduction in lithium-ion and solid-state batteries, while the Grotthuss mechanism allows for proton conduction in PEM fuel cells and flow batteries. Advanced hybrid membranes possess characteristics of both mechanisms, which include imidazole-doped polymer materials that allow proton conduction by both the vehicular and hopping mechanisms and provide continuous conduction even at low humidity or high temperatures [28]. Further enhancements include the optimization of transport using nanochannel structures that are engineered for the alignment of ionic domains for directional transport, the incorporation of ionic liquids for vehicular transport over a wide range of temperatures, and the creation of bio-inspired membranes that mimic the hydrogen bond relay mechanisms of proton channels in nature.

## 4.5 SELECTIVITY AND PERMEABILITY

One of the important issues in membrane development for electrochemical devices is the ion selectivity and permeability relationship. For instance, high permeability is desirable in electrochemical devices as it reduces the ionic resistance, enabling fast ion transport. However, high permeability reduces selectivity, causing redox species crossover, fuel leakage, or mixed ion transport. Conversely, increasing selectivity by techniques like pore size reduction or increasing charge density increases transport resistance, hence reducing efficiency [29]. The selectivity of the material is mainly determined by two factors: the pore size and the charge density. Microporous membranes that have a pore diameter comparable to or less than the solvation diameter of the ions have the ability to block the passage of larger solvated ions while allowing the passage of smaller charge carriers. For instance, in the case of VRFBs, the membranes have the responsibility of excluding the passage of vanadium ions while allowing the passage of protons and complete passage is not achieved. A similar effect has been observed in the case of PEMs, in which the sulfonic acid groups have the effect of excluding the passage of either neutral or multivalent species. In fact, the permeability-selectivity trade-off is an important phenomenon in membrane science, in that high porosity along with high pore size is associated with high permeability but low selectivity, whereas dense membranes or highly charged membranes prevent crossover but increase ionic resistance [30]. This is translated in LIBs in terms of balancing porosity to ensure conductivity, along with mechanical/thermal robustness. This is an important challenge in flow batteries, too, in that minimizing redox species crossover without incurring high ohmic losses is critical. This challenge is being addressed in recent studies by designing novel nano-transport mechanisms along with dynamic sieving phenomena, such as in graphene-based membranes that rotate. In areas where nanoscale rotational alignment is present, these membranes form channels that allow selective permeation of ions without sacrificing high permeability [31]. The anisotropic structure allows membranes to transcend the conventional trade-off boundary by allowing low resistance and high selectivity at the same time.

2D materials like graphene oxide and MXenes usually form a laminated thin membrane with an interlayer distance that matches the dimension of ions, resulting in high proton conductivity with low fuel/ion crossover. Microphase-separated domains in block copolymer membranes are used to create effective ion-conducting channels. In addition, the application of polyelectrolyte multilayers, i.e., LbL films, has been shown to control surface charge as well as ion selectivity without significantly sacrificing conductivity. Resolving the basic trade-off between permeability and selectivity is an important step in the design of electrochemical devices that can exhibit high efficiency along with long-term stability. For instance, increasing selectivity in VRFBs is known to result in high coulombic efficiency along with lower electrolyte rebalancing costs. For PEMFCs, controlling the transport of protons as well as water, along with the migration of gas crossover, is an extremely important operation. New challenges in solid-state batteries include the application of lithium-ion techniques that can inhibit dendrite growth while limiting the transport of other ions. The problems associated with membrane science that are faced because of the trade-off between permeability and selectivity are quite substantial, yet the opportunities associated with the combination of 2D materials, nanostructured channels, and hybrid composite materials are enormous. This could open the doors for the development of the next generation of membranes that are highly conductive, highly selective, and stable, thereby leading to the development of highly efficient, stable, and safe electrochemical energy storage devices.

## 4.6 MECHANICAL STRENGTH AND FLEXIBILITY

In electrochemical devices, the membranes are challenged by several harsh conditions. For instance, in a fuel cell, the membrane undergoes significant hygrothermal cycling. As a result, the membrane swells and contracts. In flow batteries, high mechanical strength is required to resist the pressure differentials. Therefore, membranes used in electrochemical devices should be able to retain their shape and structure under mechanical, chemical, or electrochemical stresses to ensure safe and durable operation. The following properties, such as tensile strength, elastic modulus, puncture resistance, and flexibility, are crucial. Mechanically weak or brittle membranes are prone to damage, deformation, and dendrite penetration, which may lead to short circuits and device failure.

### 4.6.1 Tensile Strength and Elastic Modulus

The properties of a membrane to resist mechanical stress, which result from fabrication, cell assembly, swelling, and operational cycling, must be supported by tensile strength and elastic modulus. An example of commercial polyolefin membranes is that they rely on strong tensile strength to withstand roll-to-roll processing as well as mechanical compression during battery stacking without failure. The same applies to fuel cells and electrolyzers, where ion-exchange membranes must be strong enough to endure swelling caused by hydration or exposure to acidic/alkaline environments. Many existing challenges have been addressed through reinforcement strategies. A method to enhance the strength of membranes is to incorporate nanofillers like

CNTs, graphene oxide, nanoclays, and silica nanoparticles into composite membranes, with the intention of bridging polymer chains and dissipating applied stresses to boost toughness. Ceramic-coated polyolefin membranes demonstrate increased puncture resistance and thermal dimensional stability; however, these membranes possess the surprising characteristic of maintaining their ionic transport properties [32]. In addition, the physical structure of the batteries in the form of ceramics-polymer batteries gives support to the membranes. It is also important to note that the mechanical modulus of the membranes is application-dependent. For instance, high modulus membranes are used in lithium-metal batteries. However, in the case of PEMFCs, it is important to ensure that the membranes do not exhibit high stiffness. It is important to establish the optical balance of the tensile properties, which is of vital importance in terms of mechanical strength. Additionally, flexibility is becoming an important feature in the design of wearable, flexible, and twistable energy storage devices. However, it is important to ensure that the flexible membranes exhibit the ability to twist, compress, and bend repeatedly without compromising their ionic conductivity. Electrospun nanofibrous membranes exhibit an exceptional structure that ensures that the membranes exhibit the necessary mechanical properties. Such a membrane is also capable of large deformations while maintaining the capacity for the uptake and transport of electrolytes. It is thus suitable for the development of flexible lithium-ion devices, supercapacitors, and bio-integrated devices. The recent research developments aim to design adaptive membranes that can be rigid when subjected to stress but flexible when operating normally. The design is also meant to enable the development of polymer-elastomer blends and ion-conductive hydrogels for the fabrication of flexible devices. For instance, the design of a nanocomposite membrane that utilizes the reversible hydrogen bonding interactions enables the membrane to display self-healing properties. The balance between the mechanical properties, such as strength, flexibility, durability, and toughness, is vital for membrane processing, integration, safety, and long-term durability. Stiff and brittle membranes could fail due to crack propagation, while flexible membranes could distort and even cause a short circuit. Furthermore, there is also the need to consider the long-term prospects of electrochemical devices such as flexible electronic devices, electric vehicles, and solid-state batteries with the development of robust, thermally stable, and flexible membranes.

## 4.7 CHEMICAL AND ELECTROCHEMICAL STABILITY

The fundamental need for the long-term application of the membrane in harsh electrochemical environments is the chemical and electrochemical stability of the membrane. Stability is unlike the mechanical and thermal properties of the membrane, where the long-term application is based on the resistance of the material. Stability is the ability of the membrane to withstand harsh chemical environments, including the application of high voltages and the ability of the membrane to withstand cycles without significant degradation. Electrochemical devices expose the membrane to harsh chemical environments, including exposure to radicals, acidic or alkaline environments, and oxidative environments [33]. For the case of PEMs, the main issue results from the oxidative degradation due to the presence of hydroxyl radicals (·OH) and peroxide radicals (·OOH), which are generated at the cathode due to the reduction of oxygen. To overcome this issue, stabilizing agents such as cerium oxide nanoparticles, silica, and $TiO_2$ are added to the membrane to scavenge the free radicals. On the other hand, the issue with AEMs results from the high concentration of hydroxide ions that is present in fuel cells and electrolyzers, which causes nucleophilic attack on the cationic group, such as the quaternary ammonium group, resulting in dealkylation. The solutions for these issues include the design of alkali-resistant functional groups and the combination of cross-linked or aromatic backbones that are chemically stable. Sulfide solid electrolytes exhibit high conductivity but are chemically unstable against oxide cathodes and lithium-metal anodes [34]. On the other hand, oxide electrolytes such as LLZO have better chemical stability but suffer from interfacial reactions with metallic lithium. Fillers such as silica, alumina, and titania are commonly used in this case and provide better stability.

These inorganic additives play the role of radical scavengers. They help stabilize the polymer matrices by limiting chain mobility, reducing segmental degradation during mechanical stress, and increasing solvent swell resistance. Another important requirement is that the electrochemical stability window (ESW) is sufficiently high to ensure that the membrane/electrolyte does not degrade in the voltage range. Failure to do so might result in parasitic reactions, gas evolution, and capacity loss. ESWs can extend to ≥5.3V with the incorporation of ceramics or hybrid materials. This would ensure the safe operation of future cathodes at ultra-high voltage. For PEMs and AEMs, stability is achieved by preventing gas crossover ($H_2/O_2$) and avoiding the electrochemical oxidation/reduction of functional groups. As such, the voltage range is an important indicator of stability, as is the capacity to withstand long-term high-potential operation.

In addition, the chemical and electrochemical stability of the polymer-ceramic hybrid membranes has been improved. For instance, membranes that utilize layered double hydroxides or MOFs exhibit high chemical stability against oxidative degradation and electrochemical stability against voltage breakdown, in addition to maintaining high ionic conductivity. Secondly, the application of protective coatings and interlayers at the interfaces between the electrodes and the membrane is an effective method of suppressing unwanted reactions. Overall, chemical and electrochemical stability are key factors that dictate the lifespan of the membrane. Current studies have shown that the application of radical scavengers, stable functional groups, and inorganic fillers is effective in enhancing performance in various environments.

## 4.8 THERMAL STABILITY AND HYGROSCOPIC BEHAVIOUR

Membranes in electrochemical cells experience drastic changes in temperature and hydration, and their ability to maintain structural and functional stability under such conditions is essential for long-lasting performance. The thermal stability ensures the safety, durability, and resistance to shrinkage and melting of the membranes. The hygroscopic properties and hydration-dependent ion conduction are the factors that control the water uptake for the membranes. The thermal resistance of the membranes extends the range of operation for the membranes from the maximum temperature at which no degradation, softening, or loss of functional groups occurs. The range of operation for each electrochemical cell is different, such as for PEMFCs, the PFSA membranes like Nafion are used at an operation temperature lower than 120°C. At higher temperatures, the water retention and proton conductivity are reduced. The high-temperature PEMs, like acid-doped PBI membranes, are stable at temperatures up to 200°C in dry environments [35, 36]. These membranes make the management of water simpler and improve the purity of the fuel by reducing problems such as Carbon Monoxide (CO) contamination. The transport of ions, however, can be hindered by the relaxation of polymer chains, crystallization near the ion transport regions, and mechanical instability. The incorporation of ceramic fillers such as $Al_2O_3$, $TiO_2$, etc., can improve the operation of the membranes beyond 100°C by limiting the movement of polymer segments, improving thermal conductivity, and improving the stability of the interfaces. Inorganic solid electrolytes include oxide- and sulfide-based membranes. The oxide-based membranes, such as garnet-type LLZO, are thermally stable at temperatures above 300°C. The instability of the interfaces, however, limits the performance of the membranes rather than the thermal stability. The composite membranes include fillers such as zeolites, montmorillonite clay, and MOFs, which are becoming the preferred choice over polymer matrix membranes [37]. These nanoporous fillers provide mechanical strength and thermal stability, along with the maintenance of the pathways for the conduction of ions at high temperatures. The requirement for thermal stability differs for different applications, such as low-temperature PEMFCs for automotive applications and solid-state batteries or high-temperature PEMs, for which membranes are required at temperatures over 100°C without substantial degradation.

The capacity of membranes for the absorption and retention of water (or other polar solvents) directly influences their dimensional stability and ionic conductivity. The hydration of proton membranes is critical since the conductivity of the membrane is based on the vehicle mechanism (protons are accompanied by water molecules), or the Grotthuss mechanism (proton hopping along the hydrogen-bonded water network). However, controlled water absorption increases the optimum level of hydration for proton conductivity. Excessive water absorption leads to stress, distortion, and de-lamination of the electrodes. Nafion has high conductivity at full hydration but has the disadvantage of excessive water absorption in humid environments. Improved water management through the use of fillers: Hydrophilic inorganic fillers like silica, titania, and layered clays can be used to manage water absorption in the membrane. They provide more water-sorption sites and help form water clusters. They can also limit the expansion of the polymer chains during excessive water absorption. High-temperature PEMs function in dry environments. In such environments, proton conductivity is achieved through dopants such as phosphoric acid (PA), imidazole derivatives, and ionic liquids [38]. These non-volatile proton carriers allow for conductivity even in the absence of water. The hygroscopic nature of these materials is, therefore, important in that it must be capable of absorbing sufficient water (or other proton carriers) for high conductivity, without swelling excessively, which might compromise its mechanical integrity as well as interfacial contact.

## 4.9 CONCLUSION

The membranes that have been optimized for the application of electrochemical energy systems must have a combination of properties that function in a cooperative and synergistic way, such that the long-term performance of the system will not be based on a single parameter. The performance of the electrodes in LIBs, fuel cells, electrolyzers, and RFBs is based on the combination of properties such as conductivity, selectivity, stability, and strength. Among the properties, the importance of electrical conductivity is the first and most significant. It must be such that the conductivity of the membrane is high enough, and at the same time, dimensional stability is maintained. Moreover, the strength of the membrane must be adequate so that failure of the membrane during the process of fabrication, swelling, and cycling is avoided. Nanofillers and crosslinking composites are the primary methods. Furthermore, the membrane should facilitate selective ion transport to avoid fuel leakage in fuel cells, mixing of redox species in flow batteries, and rejection of ions used in lithium- and sodium-based systems. Currently, advanced framework materials, 2D materials, and their composites hold promise in mitigating the longstanding problem of the permeability-selectivity trade-off. Moreover, stability in terms of thermal and hygroscopic properties is necessary in order to apply the membrane technology in real-world applications. It is necessary that the membrane works over a wide range of temperatures while at the same time maintaining safety, along with optimal levels of hydration or dehydrogenation carriers. Swelling too much or dehydrating/shrinking too much can cause instabilities in electrical and mechanical properties.

## REFERENCES

[1] M. J. Parnian, S. Rowshanzamir, and J. Alipour Moghaddam, "Investigation of physicochemical and electrochemical properties of recast Nafion nanocomposite membranes using different loading of zirconia nanoparticles for proton exchange membrane fuel cell applications," *Mater. Sci. Energy Technol.*, vol. 1, no. 2, pp. 146–154, 2018, doi: 10.1016/j.mset.2018.06.008

[2] A. M. Boyce, X. Lu, D. J. L. Brett, and P. R. Shearing, "Exploring the influence of porosity and thickness on lithium-ion battery electrodes using an image-based model," *J. Power Sources*, vol. 542, no. March, p. 231779, 2022, doi: 10.1016/j.jpowsour.2022.231779

[3] W. Yoopensuk et al., "Optimizing ionic conductivity and ion selectivity in zinc-polyiodide flow batteries with composite polyamide-porous separators," *J. Energy Storage*, vol. 86, no. PB, p. 111362, 2024, doi: 10.1016/j.est.2024.111362

[4] S. P. S. Badwal, S. S. Giddey, C. Munnings, A. I. Bhatt, and A. F. Hollenkamp, "Emerging electrochemical energy conversion and storage technologies," *Front. Chem.*, vol. 2, no. SEP, pp. 1–28, 2014, doi: 10.3389/fchem.2014.00079

[5] C. H. L. Tempelman, J. F. Jacobs, R. M. Balzer, and V. Degirmenci, "Membranes for all vanadium redox flow batteries," *J. Energy Storage*, vol. 32, no. 2020, p. 101754, 2020, doi: 10.1016/j.est.2020.101754

[6] F. Mokhtari et al., "Recent progress in electrospun polyvinylidene fluoride (PVDF)-based nanofibers for sustainable energy and environmental applications," *Prog. Mater. Sci.*, vol. 148, no. December 2023, p. 101376, 2025, doi: 10.1016/j.pmatsci.2024.101376

[7] M. S. Grewal, K. Kisu, S. I. Orimo, and H. Yabu, "Increasing the ionic conductivity and lithium-ion transport of photo-cross-linked polymer with hexagonal arranged porous film hybrids," *iScience*, vol. 25, no. 9, p. 104910, 2022, doi: 10.1016/j.isci.2022.104910

[8] E. B. Agyekum et al., "A state-of-the-art review of electrolyte systems for vanadium redox flow battery – status of the technology, and future research directions," *Energy Convers. Manag. X*, vol. 27, no. August, p. 101180, 2025, doi: 10.1016/j.ecmx.2025.101180
[9] M. L. Perry, J. D. Saraidaridis, and R. M. Darling, "Crossover mitigation strategies for redox-flow batteries," *Curr. Opin. Electrochem.*, vol. 21, pp. 311–318, 2020, doi: 10.1016/j.coelec.2020.03.024
[10] Y. Chen, L. Xu, X. Yang, Q. Li, M. Yao, and G. Wang, "Nanocomposite design for solid-state lithium metal batteries: Progress, challenge and prospects," *Adv. Nanocomposites*, vol. 1, no. 1, pp. 120–143, 2024, doi: 10.1016/j.adna.2024.03.002
[11] N. R. Mojarrad et al., "Advances in membranes and electrocatalysts to optimize proton-exchange membrane fuel cells," *Cell Reports Phys. Sci.*, 2025, doi: 10.1016/j.xcrp.2025.102728
[12] D. Wu, F. Xu, B. Sun, R. Fu, H. He, and K. Matyjaszewski, "Design and preparation of porous polymers," *Chem. Rev.*, vol. 112, no. 7, pp. 3959–4015, 2012, doi: 10.1021/cr200440z
[13] C. Y. Shi, W. Y. Qin, and D. H. Qu, "Semi-crystalline polymers with supramolecular synergistic interactions: From mechanical toughening to dynamic smart materials," *Chem. Sci.*, vol. 15, no. 22, pp. 8295–8310, 2024, doi: 10.1039/d4sc02089h
[14] H. Ali et al., "Impact of the nanoparticle incorporation in enhancing mechanical properties of polymers," *Results Eng.*, vol. 27, no. March, p. 106151, 2025, doi: 10.1016/j.rineng.2025.106151
[15] M. Nourani and S. M. Jafari, "Electrospun fibers/nanofibers loaded with phase change materials: Characterization, applications and challenges," *Polym. Test.*, vol. 150, no. February, p. 108879, 2025, doi: 10.1016/j.polymertesting.2025.108879
[16] T. T. K. Huynh, T. Yang, P. S. Nayanthara, Y. Yang, J. Ye, and H. Wang, *Construction of High-Performance Membranes for Vanadium Redox Flow Batteries: Challenges, Development, and Perspectives*, vol. 17, no. 1. Springer Nature Singapore, 2025, doi: 10.1007/s40820-025-01736-x
[17] Y. Zhao, H. Zhang, C. Xiao, L. Qiao, Q. Fu, and X. Li, "Highly selective charged porous membranes with improved ion conductivity," *Nano Energy*, vol. 48, no. February, pp. 353–360, 2018, doi: 10.1016/j.nanoen.2018.03.070
[18] W. Wang et al., "Hydroxide exchange membranes towards water electrolysis and fuel cells: A review on the recent advances, challenges and opportunities," *Adv. Membr.*, vol. 4, no. November, p. 100111, 2024, doi: 10.1016/j.advmem.2024.100111
[19] Z. Xue, D. He, and X. Xie, "Poly(ethylene oxide)-based electrolytes for lithium-ion batteries," *J. Mater. Chem. A*, vol. 3, no. 38, pp. 19218–19253, 2015, doi: 10.1039/c5ta03471j
[20] F. Foglia, V. Garcia Sakai, S. Lyonnard, and P. F. McMillan, "Decoupling polymer, water and ion transport dynamics in ion-selective membranes for fuel cell applications," *J. Non-Crystalline Solids X*, vol. 13, no. December 2021, p. 100073, 2022, doi: 10.1016/j.nocx.2021.100073
[21] Z. Zhu and S. J. Paddison, "Perspective: Morphology and ion transport in ion-containing polymers from multiscale modeling and simulations," *Front. Chem.*, vol. 10, no. August, pp. 1–24, 2022, doi: 10.3389/fchem.2022.981508
[22] I. A. Stenina and A. B. Yaroslavtsev, "Ionic mobility in ion-exchange membranes," *Membranes (Basel).*, vol. 11, no. 3, 2021, doi: 10.3390/membranes11030198
[23] U. Chakraborty, "Fuel crossover and internal current in proton exchange membrane fuel cell modeling," *Appl. Energy*, vol. 163, pp. 60–62, 2016, doi: 10.1016/j.apenergy.2015.11.012
[24] M. U. Javed et al., "Metal-organic framework/polymer composites for solid state electrolytes-a critical review," *J. Power Sources*, vol. 640, no. March, 2025, doi: 10.1016/j.jpowsour.2025.236720

[25] M. Inaba, T. Kinumoto, M. Kiriake, R. Umebayashi, A. Tasaka, and Z. Ogumi, "Gas crossover and membrane degradation in polymer electrolyte fuel cells," *Electrochim. Acta*, vol. 51, no. 26, pp. 5746–5753, 2006, doi: 10.1016/j.electacta.2006.03.008
[26] I. Popov et al., "Search for a Grotthuss mechanism through the observation of proton transfer," *Commun. Chem.*, vol. 6, no. 1, 2023, doi: 10.1038/s42004-023-00878-6
[27] G. Wang, J. Kang, S. Yang, M. Lu, and H. Wei, "Influence of structure construction on water uptake, swelling, and oxidation stability of proton exchange membranes," *Int. J. Hydrogen Energy*, vol. 50, pp. 279–311, 2024, doi: 10.1016/j.ijhydene.2023.08.129
[28] A. Islam et al., "Advances in filler-crosslinked membranes for hydrogen fuel cells in sustainable energy generation," *Int. J. Hydrogen Energy*, vol. 140, no. January, pp. 745–776, 2025, doi: 10.1016/j.ijhydene.2025.05.197
[29] L. Dai, K. Huang, F. Xu, S. Lin, and Z. Xu, "Emerging porous conductive ion-selective membranes for sustainable energy devices," *Angew. Chemie - Int. Ed.*, vol. 64, no. 30, 2025, doi: 10.1002/anie.202504134
[30] J. Yuan et al., "Membranes in non-aqueous redox flow battery: A review," *J. Power Sources*, vol. 500, no. February, p. 229983, 2021, doi: 10.1016/j.jpowsour.2021.229983
[31] Z. Zhang, S. Li, B. Mi, J. Wang, and J. Ding, "Surface slip on rotating graphene membrane enables the temporal selectivity that breaks the permeability-selectivity trade-off," *Sci. Adv.*, vol. 6, no. 34, pp. 1–7, 2020, doi: 10.1126/sciadv.aba9471
[32] C. Shi, P. Zhang, L. Chen, P. Yang, and J. Zhao, "Effect of a thin ceramic-coating layer on thermal and electrochemical properties of polyethylene separator for lithium-ion batteries," *J. Power Sources*, vol. 270, pp. 547–553, 2014, doi: 10.1016/j.jpowsour.2014.07.142
[33] W. Song et al., "Alkaline membranes toward electrochemical energy devices: Recent development and future perspectives," *ACS Cent. Sci.*, vol. 9, no. 8, pp. 1538–1557, 2023, doi: 10.1021/acscentsci.3c00597
[34] C. Xu, L. Chen, and F. Wu, "Unveiling the power of sulfide solid electrolytes for next-generation all-solid-state lithium batteries," *Next Mater.*, vol. 6, no. October 2024, p. 100428, 2025, doi: 10.1016/j.nxmate.2024.100428
[35] H. Chen et al., "Base-acid doped polybenzimidazole with high phosphoric acid retention for HT-PEMFC applications," *J. Memb. Sci.*, vol. 596, no. December 2019, p. 117722, 2020, doi: 10.1016/j.memsci.2019.117722
[36] L. Qingfeng, H. A. Hjuler, and N. J. Bjerrum, "Phosphoric acid doped polybenzimidazole membranes: Physiochemical characterization and fuel cell applications," *J. Appl. Electrochem.*, vol. 31, no. 7, pp. 773–779, 2001, doi: 10.1023/A:1017558523354
[37] G. M. Jaid, A. A. AbdulRazak, H. Meskher, S. Al-Saadi, and Q. F. Alsalhy, "Metal-organic frameworks (MOFs), covalent organic frameworks (COFs), and hydrogen-bonded organic frameworks (HOFs) in mixed matrix membranes," *Mater. Today Sustain.*, vol. 25, no. January, p. 100672, 2024, doi: 10.1016/j.mtsust.2024.100672
[38] A. Islam et al., "International Journal of Hydrogen Energy Recent breakthroughs and future horizons in next-generation HT-PEMs for hydrogen fuel cell," *Int. J. Hydrogen Energy*, vol. 168, no. August, p. 151113, 2025, doi: 10.1016/j.ijhydene.2025.151113

# 5 Membranes for Fuel Cells

## 5.1 INTRODUCTION

Polymer electrolyte membrane fuel cells use the membrane as both a cell separator and an electrolyte and can achieve high energy densities, which is conducive to portable applications. The primary benefits of this technology are that it eliminates the need to handle highly corrosive chemicals. The membrane plays a crucial role in conducting the ions to the cathode, where they combine with oxygen to form water [1]. The membrane's high conductivity is essential to compensate for resistive losses and ionic resistivity. Most current technologies utilize ion-conducting polymers as electrolytes in PEMFCs. PEMFCs have become the dominant fuel cell type due to their high-power density, quick response time, and versatility across various applications. The PEM membrane has three main roles, which include charge carrier for protons, to separate the reactant gases, and an electronic insulator to prevent electrons from passing through the membrane (due to having a negative charge from $SO_3^-$ and electron repelling) [2]. Nafion membrane is not suitable to be applied in fuel cells using high-concentration methanol solutions because of high methanol permeability, low proton conductivity, low water uptake, high degree of swelling, and low chemical stability at room temperature and high temperatures [3].

The PEM membrane should have several key properties for achieving higher efficiency: high proton conductivity for supporting high currents with minimal resistive losses, zero electronic conductivity, sufficient mechanical strength and stability, chemical and electrochemical stability under operating conditions, proper moisture control within the stack, very low fuel or oxygen bypass to enhance coulombic efficiency, and production costs that suit the intended application [4]. To overcome these deficiencies, various polymer-based proton-exchange membranes have been developed. Some of the polymer-based materials developed to overcome the deficiencies of existing PEMs are: cellulose-based materials and its derivatives, PBI, PFSA, poly(ether ether ketones) (PEEK), polystyrene sulfonate (PSS), PVDF, etc. [5]. Composite membranes are among those that contain nontoxic, highly hydrophilic organic and inorganic nanoparticles, which, with the help of polymers, sustain a uniform structure to boost proton conductivity and provide coverage to minimize methanol crossover. Different types of nanomaterials, including $Al_2O_3$, $SiO_2$, $TiO_2$, and GO, have been identified as the most promising candidates for composite membrane applications. The nanoparticle-added membrane not only stabilizes the Grotthuss and vehicular mechanisms through strengthening the channel walls and providing hydrophilic groups, but it also increases water retention capacity, which in turn facilitates high proton conduction and fuel cell stability.

DOI: 10.1201/9781003679363-5

The performance of the PEM depends on its proton conductivity and humidity levels, and increased humidity enhances proton conductivity. One way to reduce water drag or crossover is by decreasing the membrane thickness, which can enhance fuel cell performance. The high cost of PEMs remains a critical challenge, emphasizing the importance of developing economically viable and scalable fabrication approaches [6]. Future advancements in PEMs are moving toward multifunctional membranes that conduct protons while offering self-healing, radical-scavenging, and adaptive hydration properties. Emerging advances in nanotechnology and bio-inspired hybrid architectures are paving the way for a new generation of durable, efficient, and low-cost fuel cell membranes. This chapter mainly discusses fuel cell membrane materials, their composition, structure–property relationships, operational challenges, and strategies for improving performance.

## 5.2 MEMBRANE TYPES AND MATERIALS

### 5.2.1 Perfluorinated Sulfonic Acid Membranes

PFSA membranes are commonly used for fuel cells and other electrochemical devices due to their superior proton conductivity and chemical stability. The poly-(tetrafluoroethylene-co-perfluorosulfonic acid) based polymer is the most commonly used PEM for fuel cells [7]. The hydrophobic PTFE backbone imparts mechanical strength, while the hydrophilic side chains enable high proton conductivity under fully hydrated conditions. The hydrated membrane exhibits enhanced proton conductivity due to improved percolation pathways in the hydrophilic regions. In the hydrated Nafion membrane, mobile protons readily associate with water molecules to form $H_3O^+$ complexes. This generates hydronium ions, which promote the effective transport of protons across the membrane's sulfonic acid ($SO_3H$) groups, which are essential for efficient proton conduction. This nanoscale separation between hydrophilic sulfonic acid groups ($-SO_3H$) and the hydrophobic fluorocarbon backbone is critical for establishing interconnected proton-conducting pathways. Although various methods were utilized to enhance the properties of the Nafion membrane, a promising method to enhance the properties of the Nafion membrane is to increase its proton conductivity by adding ionic groups. Nevertheless, it is difficult to incorporate ionic groups into the Nafion membrane due to its poor mechanical properties, which are a result of its swelling characteristics. Extensive efforts have been devoted to the modification of Nafion membranes using inorganic particles [8]. Various inorganic particles, such as metal oxide, heteropolyacids, metal phosphate, clay, zeolite, graphene oxide, and many others, were utilized to develop inorganic-Nafion composite membranes. Recent advancements in this field focus on incorporating inorganic fillers (such as $SiO_2$, $TiO_2$, $ZrO_2$, and graphene oxide) to improve water retention and radical resistance [9], as well as creating reinforced and multilayer structures to control swelling and shrinkage. incorporating 5 wt.% $SiO_2$, the durability of the modified Nafion membrane increased to 50 hours at an operating temperature of 130°C without performance loss. In their research, the authors prepared a composite membrane of Nafion with polyoxometalate coupled with graphene oxide, which acts as a linker for PEMFCs. To enhance the proton conductivity of the Nafion membrane, the authors used phosphotungstic acid, which acts as the linker with the GO, and the results showed improved conductivity compared to the recast membrane. The Nafion composite membrane outperformed the pure Nafion membrane in terms of fuel cell performance.

### 5.2.2 Hydrocarbon-Based Membranes

Hydrocarbon-based membranes for fuel cells are gaining interest because they potentially facilitate the recycling of membrane-electrode assemblies (MEAs) as they do not contain fluorinated compounds harmful to the environment. Hydrocarbon-based PEMs consist of aromatic polymers functionalized with sulfonic acid (-$SO_3H$) groups on either the main chain or the side chains, as shown in Figure 5.1 [10]. Some of the examples of hydrocarbon-based PEMs are: SPEEK, polyphenylenesulfones (sPPS), PES, and polystyrenes (sPS). It has been found that these materials have superior thermal stability and gas barrier properties compared to existing commercial materials based on PFSA. It has also been found that hydrocarbon-based polymers would require less expensive starting materials and would be much simpler to produce, thus reducing costs. The hydrocarbon polymers offer potential for use at elevated temperatures (above 100°C) owing to the high thermomechanical stability of the underlying polymers used in hydrocarbon membranes, e.g., polyphenylene, polyphenylene sulfone, and poly(arylene ether sulfone) [11]. Due to the higher tortuosity of the hydrophilic phase and lower solubility of oxygen in the hydrophobic phase, which may be related to the less pronounced hydrophilic-hydrophobic phase separation in hydrocarbon membranes, the gas permeability of hydrocarbon membranes is often lower compared to that of membranes based on PFSA. To enhance the mechanical

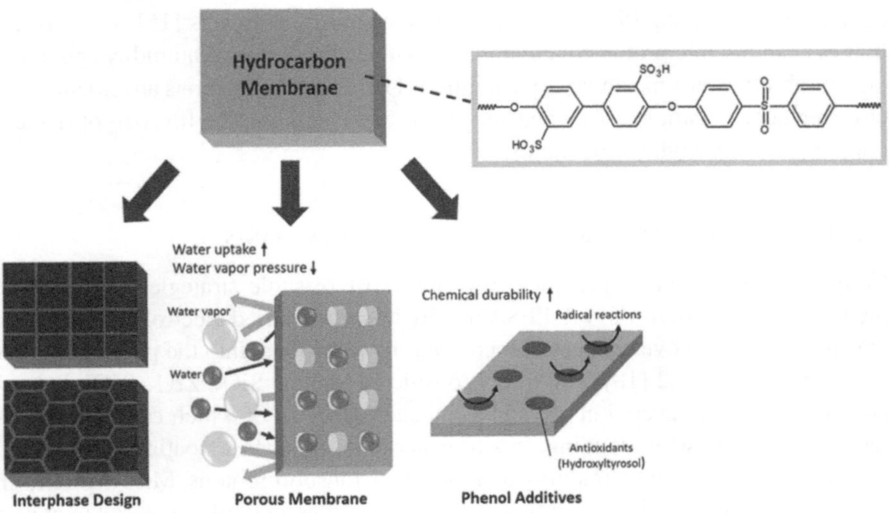


**FIGURE 5.1** Schematic representation of performance enhancement strategies in hydrocarbon-based proton-exchange membranes.

and chemical stability, promising strategies identified in PFSA-based membranes have also been considered for hydrocarbon-based membranes. For example, inert substrates, hydroxyl radical scavengers (such as $CeO_2$ nanoparticles), and nanofibers can be incorporated into hydrocarbon-based membranes. There is interest in developing nanostructured hydrocarbon copolymers that mimic the microphase-separated structure of PFSA-based membranes, thereby providing lower-cost, more robust membranes.

### 5.2.3 High-Temperature Membranes

High-temperature proton-exchange membranes (HT-PEMs) have been developed to overcome the limitations of PFSA electrolytes, especially their dependence on external humidification and their limited operating temperature range (<90°C) [12]. Among these, phosphoric acid-doped PBI membranes are particularly significant. PBI is a high-performance aromatic heterocyclic polymer with a rigid backbone that facilitates substantial phosphoric acid uptake via acid–base complexation with its imidazole groups. A phosphoric acid-doped PBI membrane can achieve good conductivity and operate stably between 140 and 200°C without external humidification [13]. However, several issues limit the full utilization of its beneficial properties. For example, acid leaching is a major concern, as phosphoric acid gradually escapes from the membrane, causing long-term loss of conductivity and performance declines [14]. High acid loadings can weaken the mechanical structure because the swollen, porous polymer backbone becomes fragile, leading to deformation. A major challenge in developing materials is balancing increased acid uptake for better conductivity with maintaining dimensional and mechanical stability. To tackle these issues, current approaches include crosslinking PBI, adding nanofillers such as silica, titania, or graphene oxide to lock in phosphoric acid, and creating membrane blends or composites combining PBI with other high-temperature polymers [15]. These membranes enable stable proton conduction without needing external humidity and maintain durability over time under dry conditions [16]. Such innovations are essential for ensuring acid retention, enhancing longevity, and extending the lifespan of devices operating at very high temperatures.

### 5.2.4 Composite and Inorganic-Enhanced Membranes

Composite membranes represent one of the most versatile strategies to overcome the inherent limitations of both PFSA and hydrocarbon-based electrolytes, based on the simple yet effective concept of incorporating nanofillers into the polymer matrix as given in Figure 5.2 [18]. The fillers considered include $SiO_2$, $ZrO_2$, $TiO_2$, heteropolyacids (HPAs), layered clays, and graphene oxide, all of which can be incorporated through physical blending, in situ generation, or surface coating as shown in Figure 5.2 [19]. Due to their ability to adsorb and transport protons, $MO_2$ (M = Zr, Si, Ti, Sn, Si) nanoparticles have shown promise as an inorganic filler material in hybrid membranes. The conductivity of hydrated protonic oxide increases with an increase

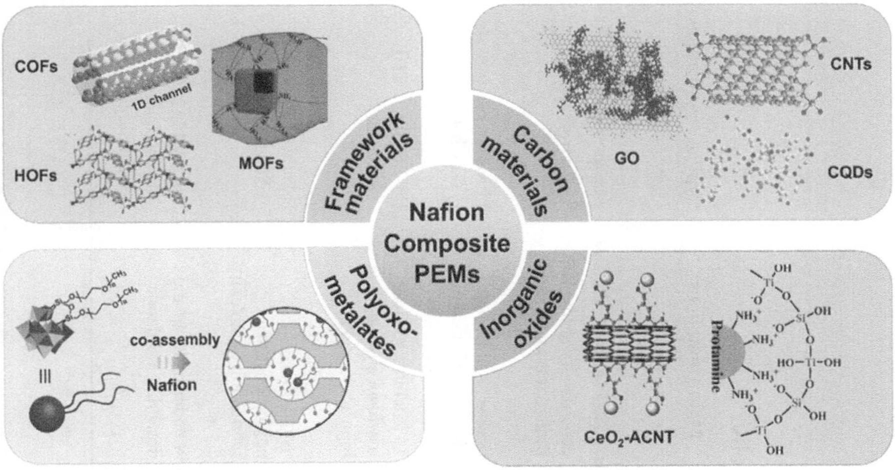


**FIGURE 5.2** Overview of common nanofillers incorporated into composite proton-exchange membranes.

**(Reproduced with permission [17].)**

in hydration [19]. Various studies have reported that the protonic conductivity of $2.3 \times 10^{-2}$ S $cm^{-1}$ for $SnO_2.nH_2O$ at 150°C under saturated water vapour pressure. The surface acidity of metal oxides (PIE) is known to affect the proton conductivity of membranes. For example, it has been reported that $SnO_2$ has higher proton conductivity than $ZrO_2$, as Sn has higher electronegativity than Zr. Hence, it is logical to conclude that composite membranes using $SnO_2$ as an inorganic filler material have higher conductivity than those using $TiO_2$, as Sn has higher electronegativity than Ti. The difference in conductivity can be related to the PIE number as well. However, it is also affected by morphology, crystallinity, and size. Different varieties of metal-organic frameworks, such as ZIF-8, HKUST-1, and UiO-66, are synthesized and utilized as PEMs in various fuel cells.

To improve the composite membrane's performance, it is important to functionalize the membrane and introduce sulfonic groups. Sulfonic groups are known to offer significant water retention ability, thus making it suitable for DMFC applications. Composite-based membrane is prepared by co-doping functional groups of MOFs containing both $SO_3H$ and $SO_3$–$NH_2$ groups with a solution of Nafion. The two functionalized MOF-based composite membranes created proper ionic channels, and the amino and sulfonic groups bound to more water molecules, which promoted proton conduction. The value of proton conductivity of the composite membrane reached $256 \times 10^{-3}$ S $cm^{-1}$ (90°C and 95% RH), which is almost 1.17 times greater than that of the recast Nafion membrane ($118 \times 10^{-3}$ S $cm^{-1}$). Table 5.1 offers a brief overview of different membrane types by summarizing their key materials, temperature ranges, proton conductivity, benefits, drawbacks, and main uses.

**TABLE 5.1**
**Comparison of Representative Proton-exchange Membrane (PEM) Types Used in Fuel Cells**

| Membrane Type | Representative Materials | Operating Temperature (°C) | Proton Conductivity (S cm$^{-1}$) | Key Advantages | Main Limitations | Typical Applications |
|---|---|---|---|---|---|---|
| **PFSA** | Nafion®, Aquivion® | 60–90 | 0.10–0.20 (hydrated) | High conductivity, good chemical/mechanical stability, mature technology | High cost, poor conductivity at low RH, limited thermal stability | PEMFCs, PEM electrolyzers |
| **Hydrocarbon-based membranes** | SPEEK, SPPSU, SPSU, polyarylene ethers | 80–120 | 0.10–0.18 | Low-cost, thermally stable, scalable fabrication | Oxidative degradation, dimensional instability at high sulfonation | PEMFCs, electrolysis, flow batteries |
| **High-temperature membranes** | $H_3PO_4$-doped PBI | 140–200 | 0.05–0.10 (anhydrous) | Operates without humidification, high CO tolerance, stable at high T | Acid leaching, reduced mechanical strength at high acid uptake | HT-PEMFCs, CHP systems |
| **Composite/ Inorganic-enhanced membranes** | Nafion-$TiO_2$, Nafion-$ZrO_2$, SPEEK-$SiO_2$, GO-based hybrids | 80–140 | 0.05–0.12 | Improved water retention, radical resistance, and dimensional stability | Possible filler agglomeration, fabrication complexity | Advanced PEMFCs, electrolyzers |
| **Fluorine-free next-generation PEMs** | Pemion®, sulfonated polyphenylene | 60–150 | 0.08–0.15 | Eco-friendly, cost-effective, high stability | Still under scale-up and durability validation | Sustainable PEMFCs, future hydrogen technologies |

## 5.3 ROLES AND KEY CHARACTERISTICS OF THE MEMBRANE OF PEMS

PEMs are the core component of a PEMFC and a proton-exchange membrane electrolyzer. Their main role is to selectively transport protons from the anode to the cathode. Because of its dual function, the electrochemical reaction proceeds efficiently: electrons are restricted to the external circuit, and gas passage through the membrane for recombination is kept minimal [7]. This helps reduce parasitic losses, conserves fuel, and enhances safety. The properties that define the PEM's performance determine its overall effectiveness. An overview of the PEMFC structure, catalyst layers, and performance–reliability relationship is illustrated in Figure 5.3.

### 5.3.1 Proton Conductivity

The proton conductivity of these membranes is related to the hydration of the membrane and is linked to the vehicle mechanism with sufficient carriers in the membrane and the Grotthuss hopping mechanism without sufficient carriers. The transition from one mechanism to another is characterized by a clear change in activation energy

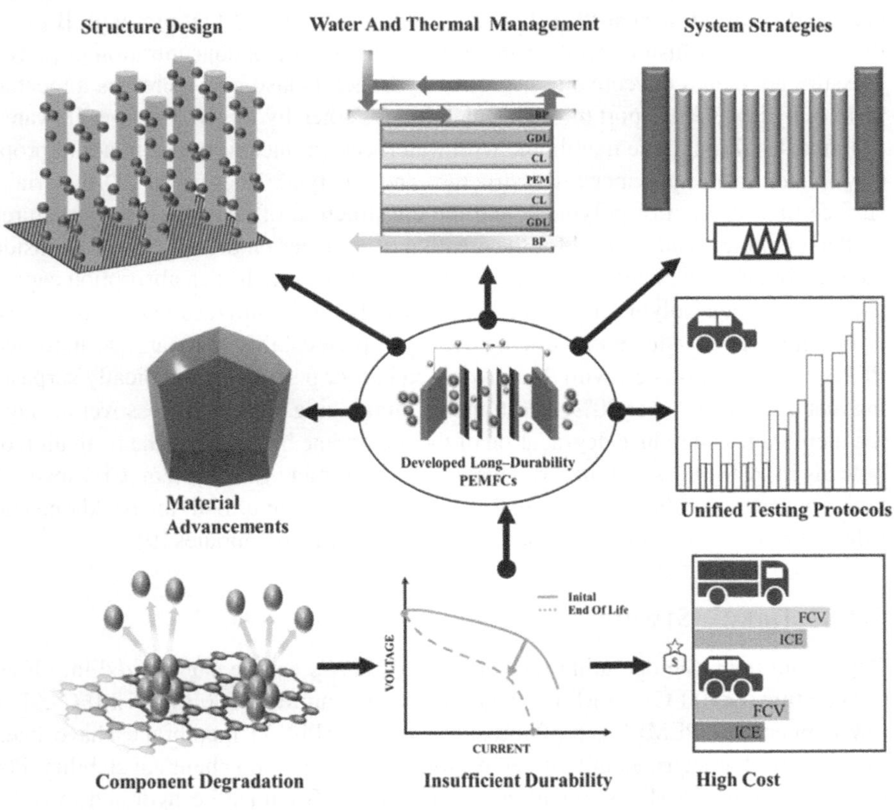


**FIGURE 5.3** Schematic illustration of major PEMFC components.

($E_a$(sd) > $E_a$(vd)). Another parameter to consider when describing proton conductivity in these membranes is the dielectric constant of the polymer and the vehicle. The conductivity of some sulfonated membranes decreases in the presence of methanol because its lower dielectric constant than water increases ion pairing, thereby decreasing dissociation. For the PEMFC system, it is desirable to have as high ionic conductivity as possible (50 mS $cm^{-1}$ at ambient temperature), which is stable over time and in fuel cell conditions (e.g., different humidity conditions and different partial pressures of oxygen $P(O_2)$)). For proton transport in PEMs, two parameters exist: the ion exchange capacity (IEC) and $\lambda$. Sulfonated polystyrene is also considered to be a significant contender for PEMFC and is being researched extensively. The sulphonated membranes with nano-silica had shown proton conductivity of 0.432 $S{\cdot}cm^{-1}$ and IEC of 3.21 mmol. $g^{-1}$, which is higher than that of Nafion 117, which is 0.374 S·cm−1 and 0.94 mmol·g−1, along with a reduction of 35% in methanol crossover [20].

### 5.3.2 Low Gas Permeability

Gas permeation through the membrane of PFSA is a dissolution-diffusion process. In this process, gas molecules dissolve at one side of the membrane at a high pressure, diffuse through the membrane due to the difference in concentration gradients, and finally desorb at the other side of the membrane at low pressure. Besides the dissolution-diffusion mechanism based on the hydrogen concentration or partial pressure gradient, convective transport of hydrogen is also considered as a mechanism of hydrogen transport through a membrane. When hydrogen dissolves in water, it permeates through the membrane with water. Gas permeability is a material property related to the polymer microstructure and gas type. The use of 2D materials, surface modification of polymers, and the construction of layer-by-layer structures on the membrane can be used for the creation of a barrier on the membrane or inside the membrane, which has low hydrogen permeability or hydrogen absorption capacity. Hydrocarbon-polymer-based PEMs, especially those with designable structures, offer a notable advantage in reducing hydrogen permeability compared to modified PFSA-based membranes, with their hydrogen barrier performance typically surpassing that of PFSA by over 50%. Additionally, minimizing hydrogen crossover through the membrane can reduce degradation of the membrane by reducing the formation of radicals in PEMFC, which in turn increases the lifespan of the system. Crossover of hydrogen creates radicals, which in turn cause degradation of both the backbone and side chains of the membranes, leading to defects in the membranes [9].

### 5.3.3 Thermal Stability

Depending upon the operating temperature, PEMFCs can be categorized into low-temperature (<100°C) PEMFCs and high-temperature (>100°C) PEMFCs. For low-temperature PEMFCs, perfluorosulfonic acid (PFSA) membranes have been extensively characterized in terms of proton conductivity and chemical stability. For low-temperature PEMFCs, membranes based on PFSA must be hydrated in order to have high proton conductivity; however, this is related to the level of hydration.

Various parameters have a significant impact on PEMFC performance. These parameters include the characteristics of polymer electrolytes, temperature, humidity, catalysts, and loading percentage. In low-temperature PEMFCs, Nafion and composite membranes made from Nafion have been researched as PEMs owing to their good proton conductivity at humidified conditions and good mechanical properties. However, Nafion exhibits reduced proton conductivity above 100°C due to membrane dehydration and the consequent deterioration of its chemical and mechanical stability. Hybrid PEM has been synthesized with -OH functionalized ionic liquids, nano-silica, and Nafion for fuel cell applications according to a research study. PEMs hybrid membrane exhibited good thermal stability, good mechanical properties, and good proton conductivity compared to other PEMs made from ionic liquids [21]. The Nafion composite membrane, comprising IL-OH and silica, exhibited good proton conductivity (~55 mS cm−1) above 160°C. The fuel cell made of this novel PEM produced power densities of 420 mW $cm^{-2}$ at 180°C. The proton conductivity of polymer/IL composite membranes was observed to be insufficient for commercial applications due to the high viscosity of ILs.

### 5.3.4 Chemical Durability

The PEM plays a crucial role in facilitating the transport of protons between the anode and cathode while preventing $H_2$ and $O_2$ from coming into contact with each other. Both types of membranes undergo chemical degradation over long periods of use. This degradation is attributed to various kinds of attacks by radicals. These radicals include hydroxyl radicals, hydrogen radicals, and hydroperoxyl radicals. They weaken the membrane structure, and this process may be accelerated by metal contaminants. From the experiment, it was noted that the reinforced membrane showed better conductivity and improved chemical and mechanical durability under extreme conditions of open circuit voltage hold and humidity cycle tests. There is a possibility of composite membranes of Nafion and $SiO_2$, $ZrO_2$, $TiO_2$, and graphene oxide showing better chemical and mechanical durability. Moreover, there has been significant interest in developing membranes with chemical stability against free radicals. The incorporation of free radical scavengers and hydrogen peroxide decomposition catalysts, such as cerium and manganese compounds, has shown promise in preventing membrane degradation. The membrane supported on $CeO_2$ had a lifespan six times longer and an emission rate 40 times lower than the baseline.

## 5.4 PERFORMANCE LIMITATIONS OF THE FUEL CELL

### 5.4.1 Ionic Conductivity and Water Retention

The major chemical, mechanical, and thermal degradation pathways in PEMFC membranes are summarized in Figure 5.4. The presence and distribution of water within the membrane are essential factors influencing proton conductivity in PEMs. In typical PFSA systems such as Nafion, conduction primarily occurs via a vehicular mechanism, in which protons travel with water molecules, and a Grotthuss hopping mechanism, which relies on a continuous hydrogen-bonded water network.

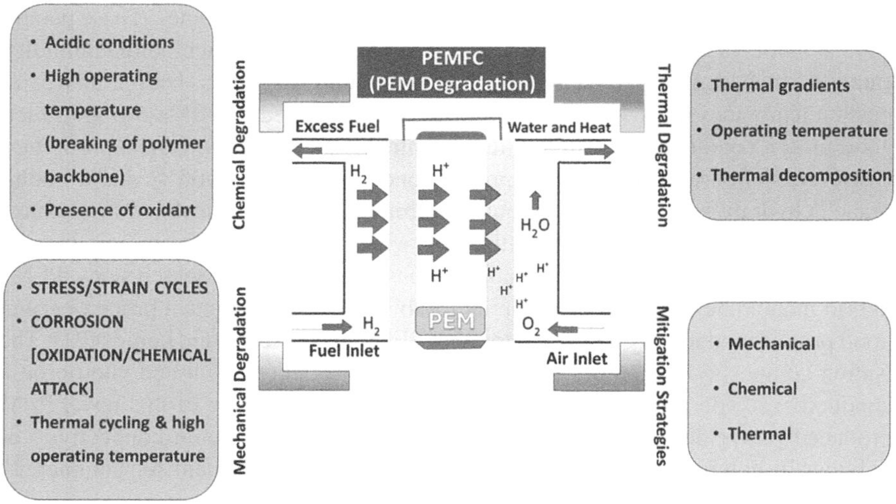


**FIGURE 5.4** Overview of PEM degradation mechanisms, including chemical, mechanical, and thermal degradation pathways [22].

The decrease in conductivity at low humidity or high temperatures presents one of the biggest challenges during extended PEMFC operation. Insufficient membrane hydration not only increases ohmic resistance but also accelerates membrane degradation, as dehydrated backbones become more vulnerable to chemical and mechanical stress. This issue is compounded by unavoidable factors, including thermal cycling and fluctuating humidity, in automotive PEMFCs. The problems with hydration can be handled with two methodologies: external and internal humidification. In the external humidification method, the control of the humidity levels of the reactant gases is controlled by adjusting the temperature of the humidification and the time of interaction between the reactant gases and water in the external humidification chamber. The internal humidification method involves the direct addition of water into the PEMFC. This has two objectives: the hydration of the membrane and the maintenance of the moisture content at desired levels. The voltage levels are maintained at high levels when the water management system is integrated, implying that the water management has been optimized, and the membrane degradation has been mitigated [23]. These approaches collectively work to support efficient ionic transport in real-world situations, such as PEMFCs, by minimizing dependence on bulk water while preserving membrane conductivity.

### 5.4.2 Durability and Degradation

The chemical decomposition of PEMs is generally attributed to factors such as hydrogen peroxide formation, reactant gas crossover, platinum catalyst dissolution, recrystallization, and the presence of cationic contaminants. The initiation of localized degradation of membranes is often observed at the cathode side and is enhanced in dry and high-temperature environments. An often-employed criterion for detecting degradation in fluorine membranes is the presence of hydrogen fluoride and/or chain segments in condensed discharge water. In PEM fuel cells, oxygen molecules may permeate through the polymer membrane. This is often observed when there is a defect or imperfection in the membrane. Once the oxygen molecules come into contact with the anode side, incomplete reduction may occur. This leads to the formation of highly reactive free radicals like •OH and $HO_2$•. Incomplete reduction is defined as the reduction of oxygen molecules to water in electrochemical reactions at the anode side [20]. As a result, active radicals are formed on the catalyst surface. Such active radicals include $HO^{\cdot}$ radicals. In summary, it is clear that the rate of membrane degradation is directly proportional to the amount of gas crossover, where higher amounts of oxygen are present in the anode. Such an incomplete reduction reaction is also indicated by the presence of hydrogen peroxide ($H_2O_2$) in the water produced. $H_2O_2$ is a by-product of the oxygen reduction reaction (ORR). Hence, it is an indicator of active radicals in a reaction. An increase in gas crossover results in a higher intensity of this incomplete reduction reaction, leading to a higher amount of active radicals and thus higher rates of membrane degradation. Such problems can be reduced to a certain extent by changing the polymer structure and adding fillers such as metal oxide or carbon nanotubes to it, or using non-fluorine-based membranes and polymers with shorter side chains. Another important aspect of PEMFCs is to ensure that the membrane is humidified to a certain level. This can be done by passing gases through external humidifiers maintained at a required temperature. Also, it is essential to maintain uniform membrane thickness to effectively manage mechanical failures. The latest PEM designs focus on stabilizing polymer backbones, synthesizing radical-scavenging composites, and reinforcing structures to extend the lifespan of fuel cells beyond current industry goals.

### 5.4.3 Cost and Scalability

The economic viability of fuel cell technologies primarily relies on the cost and scalability of proton-exchange membranes. Nafion and other PFSA membranes, known for their high performance, are still costly mainly because of their complex fluoropolymer synthesis, with production costs spanning approximately $600 to $1,200 per square metre [9]. This high cost represents a major portion of the overall stack expense, restricting their application in key sectors like automotive and stationary power. To tackle this challenge, research is mainly focused on creating non-fluorinated alternatives that are more cost-effective yet retain performance. Hydrocarbon-based proton-exchange membranes like SPEEK, sulfonated polysulfone, and degradable block copolymers can be synthesized on a large scale from inexpensive raw materials. Importantly, commercial options such as Pemion show that non-fluorinated membranes can compete with, or even outperform, PFSA-based systems in terms of cost and electrochemical performance. Composite and hybrid membrane systems offer an exciting path forward, using affordable hydrocarbon backbones that are reinforced or functionalized with nanofillers to match the properties of PFSA membranes [21]. Recent advancements in high-temperature PBI membranes emphasize their potential as fluorine-free options, removing the need for costly perfluorinated chemistries. They can function at high temperatures without external humidification, which simplifies water management and reduces overall system costs. Hydrocarbon membranes and composites are more environmentally friendly and scalable for large-scale polymer processing methods like extrusion, roll-to-roll casting, and 3D printing, enabling straightforward routes to mass production.

## 5.5 FUTURE PERSPECTIVES ON PEM

The durability of PEMFCs is another key issue that is preventing the commercialization of PEMFCs for automotive and stationary power generation applications. For example, it is necessary to upgrade PEMFC systems in order to extend operating hours in the future. Therefore, it is apparent that the commercialization of hydrogen-based cells is a major challenge due to failures and degradation in various parts of the cell system. For instance, the accelerators of PEM degradation include high partial pressures of reactants, low RH levels, high-temperature levels, and high voltage levels. Therefore, it is crucial to address all of these factors in order to mitigate degradation and maximize fuel cell longevity. An emerging PEM is made of composite membranes comprising polymers and inorganic nanofillers. These membranes have particles that are tightly bound in the polymer matrix with multiple ionic channels and free volume for holding water molecules. However, gas crossover, especially of hydrogen and oxygen, should be reduced to a minimum to ensure safety and maximum efficiency. This can be done using nanocomposite membranes with impermeable particles such as clay and graphene, which are non-permeable to gases while still being permeable to protons. Another significant factor is that it is important to ensure that the membrane is chemically stable to prevent it from deteriorating in oxidative and acidic environments. This would help to increase its life expectancy. Cost-effective materials are also important in this case, possibly in the form of cheaper raw materials or easier manufacturing processes. Using nontoxic and halogen-free chemicals would also help to reduce environmental impact and increase safety in the long run. Despite all these benefits, however, there are still problems with the composite membrane in that it is not performing well in terms of proton conduction and methanol or fuel permeability/crossover. To solve this problem, there is a pressing need to develop novel proton-conducting membranes/components that are able to perform in higher-temperature applications.

## 5.6 CONCLUSION

Proton-exchange membranes (PEMs) are a critical component of fuel cell systems, requiring materials that combine efficient ion conduction, long-term durability, and economic viability. Among the main material categories, perfluorosulfonic acid membranes, hydrocarbon-based ionomers, high-temperature phosphoric-acid-doped PBI membranes, and composite or inorganic-enhanced systems each represent different points on a spectrum of strength and trade-offs. These technologies still face challenges such as continuous proton conduction at high temperatures without external humidification, long-term chemical and mechanical durability, prevention of gas crossover and catalyst contamination, and development of cost-effective, fluorine-free alternatives to traditional PFSA systems. Recent advances in membrane engineering that address these issues include multilayer reinforcement, nanoparticle incorporation, targeted additive strategies, and the design of new, more affordable, and environmentally friendly hydrocarbon ionomers. These developments point toward the next generation of high-performance, durable, and market-ready PEMs, which are crucial for fuel cells that serve as clean energy conversion and storage systems and represent a key component of highly decarbonized energy infrastructures.

## REFERENCES

[1] N. R. Mojarrad et al., "Advances in membranes and electrocatalysts to optimize proton-exchange membrane fuel cells," *Cell Reports Phys. Sci.*, 2025, doi: 10.1016/j.xcrp.2025.102728

[2] M. Kiani, Y. Zhao, and R. Zhang, "Proton exchange membrane fuel cells: Recent developments and future perspectives," *Chem. Commun.*, vol. 61, no. 52, pp. 9392–9411, 2025, doi: 10.1039/d5cc01478f

[3] E. Wallnöfer-Ogris et al., "A review on understanding and identifying degradation mechanisms in PEM water electrolysis cells: Insights for stack application, development, and research," *Int. J. Hydrogen Energy*, vol. 65, no. December 2023, pp. 381–397, 2024, doi: 10.1016/j.ijhydene.2024.04.017

[4] S. H. Mirfarsi et al., "High-temperature stability of hydrocarbon-based Pemion® proton exchange membranes: A thermo-mechanical stability study," *Int. J. Hydrogen Energy*, vol. 50, pp. 1507–1522, 2024, doi: 10.1016/j.ijhydene.2023.07.236

[5] A. Islam et al., "Recent breakthroughs and future horizons in next-generation HT-PEMs for hydrogen fuel cell," *Int. J. Hydrogen Energy*, vol. 168, no. August, p. 151113, 2025, doi: 10.1016/j.ijhydene.2025.151113

[6] S. Ahmed et al., *Recent Advancements in Designing High-performance Proton Exchange Membrane Fuel Cells: A Comprehensive Review*, vol. 390, no. March. Elsevier Ltd, 2025. doi: 10.1016/j.apenergy.2025.125753

[7] C. Zhou, M. A. Guerra, Z.-M. Qiu, T. A. Zawodzinski, and D. A. Schiraldi, "Chemical durability studies of perfluorinated sulfonic acid polymers and model compounds under mimic fuel cell conditions," *Macromolecules*, vol. 40, no. 24, pp. 8695–8707, Nov. 2007, doi: 10.1021/ma071603z

[8] S. Samantaray, D. Mohanty, S. K. Satpathy, and I. M. Hung, "Exploring recent developments in graphene-based cathode materials for fuel cell applications: A comprehensive overview," *Molecules*, vol. 29, no. 12, 2024, doi: 10.3390/molecules29122937

[9] E. Y. Safronova, A. A. Lysova, D. Y. Voropaeva, and A. B. Yaroslavtsev, "Approaches to the modification of perfluorosulfonic acid membranes," *Membranes (Basel).*, vol. 13, no. 8, 2023, doi: 10.3390/membranes13080721

[10] H. Nguyen et al., "Hydrocarbon-based PemionTM proton exchange membrane fuel cells with state-of-the-art performance," *Sustain. Energy Fuels*, vol. 5, no. 14, pp. 3687–3699, 2021, doi: 10.1039/d1se00556a

[11] J. Song et al., "Rational materials and structure design for improving the performance and durability of high temperature proton exchange membranes (HT-PEMs)," *Adv. Sci.*, vol. 10, no. 30, pp. 1–26, 2023, doi: 10.1002/advs.202303969

[12] K. A. Perry, K. L. More, E. A. Payzant, R. A. Meisner, B. G. Sumpter, and B. C. Benicewicz, "A comparative study of phosphoric acid-doped m-PBI membranes," *J. Polym. Sci. B Polym. Phys.*, vol. 52, no. 1, pp. 26–35, 2014, doi: 10.1002/polb.23403

[13] Z. Xu et al., "Strategies for mitigating phosphoric acid leaching in high-temperature proton exchange membrane fuel cells," *Molecules*, vol. 29, no. 18, 2024, doi: 10.3390/molecules29184480

[14] E. Quartarone, P. Mustarelli, A. Carollo, S. Grandi, A. Magistris, and C. Gerbaldi, "PBI composite and nanocomposite membranes for PEMFCs: The role of the filler," *Fuel Cells*, vol. 9, no. 3, pp. 231–236, 2009, doi: 10.1002/fuce.200800145

[15] J. Li et al., "Porous polybenzimidazole membranes doped with phosphoric acid: Preparation and application in high-temperature proton-exchange-membrane fuel cells," *Energy Convers. Manag.*, vol. 85, pp. 323–327, 2014, doi: 10.1016/j.enconman.2014.05.052

[16] T. S. Mayadevi et al., "Nafion composite membranes impregnated with polydopamine and poly(sulfonated dopamine) for high-performance proton exchange membranes," *ACS Omega*, vol. 7, no. 15, pp. 12956–12970, 2022, doi: 10.1021/acsomega.2c00263

[17] S. Song, H. He, S. Chai, and H. Li. "Advanced Nafion/nanofiller composite proton exchange membranes for fuel cell applications," *Polymer.*, vol. 307, p. 127241, 24 Jul. 2024.

[18] M. Tanahashi, "Development of fabrication methods of filler/polymer nanocomposites: With focus on simple melt-compounding-based approach without surface modification of nanofillers," *Materials*, vol. 3, no. 3, pp. 1593–1619, 2010, doi: 10.3390/ma3031593

[19] M. B. Patil, S. L. Bhagat, R. S. Sapkal, and V. S. Sapkal, "A review on the fuel cells development," *Sci. Revs. Chem. Commun*, vol. 1, no. 1, pp. 25–41, 2011.

[20] X. Luo, L. Ghassemzadeh, and S. Holdcroft, "Effect of free radical-induced degradation on water permeation through PFSA ionomer membranes," *Int. J. Hydrogen Energy*, vol. 40, no. 46, pp. 16714–16723, 2015, doi: 10.1016/j.ijhydene.2015.07.118

[21] S. H. Woo et al., "Functionalized nanoclays for improved properties of composite proton exchange membranes," *J. Power Sources*, vol. 549, no. March, p. 232083, 2022, doi: 10.1016/j.jpowsour.2022.232083

[22] T. Sharma, U. Adhikari, A. Nandimath, and J. Pandey. Investigating degradation & mitigation strategies for proton conducting membrane in proton exchange membrane fuel cell: An approach to develop an active & stable membrane. *Mater. Today Sustain.*, vol. 30, p. 101103, 1 Jun. 2025.

[23] J. Kerres and V. Atanasov, "Cross-linked PBI-based high-temperature membranes: Stability, conductivity and fuel cell performance," *Int. J. Hydrogen Energy*, vol. 40, no. 42, pp. 14723–14735, 2015, doi: 10.1016/j.ijhydene.2015.08.054

# 6 Membranes for Supercapacitors

## 6.1 INTRODUCTION

Supercapacitors, also known as electrochemical capacitors, are electrochemical energy storage devices that bridge the performance gap between conventional dielectric capacitors and rechargeable batteries [1]. Batteries store energy primarily through redox reactions involving electron transfer (faradaic processes). In contrast, supercapacitors rely on electrostatic mechanisms, especially at the interface between the electrodes and the electrolyte (electric double layer capacitors, EDLCs), or on rapid surface redox reactions. This mechanism enables supercapacitors to offer high-power density, very short discharge/charge time, and high cycle life that can extend to millions of cycles [2]. As a result, they are widely used in regenerative braking systems, power backup modules, portable and flexible electronic devices, among others. The membrane plays a pivotal role in supercapacitor operation and is often overlooked in assessments of supercapacitor efficiency and long-term stability. The membrane has two major functions: (i) it serves as an ion-permeable electrolyte reservoir, allowing rapid ion transport between the electrodes, and (ii) it serves as an insulator, which prevents short circuiting while providing structural integrity during operational cycling [3]. Physicochemical properties of the membrane, such as thickness, porosity, ionic conductivity, wettability, and thermal and mechanical stability, have a direct influence on the safety, power, and reliability of the device. This chapter will cover the membrane functions in the context of material selection, from traditional polyolefin membranes through advanced polymer composite materials, fibrous composite materials, and even solid-state polymer electrolytes, based on the basic understanding of membrane fundamentals. It also examines new fabrication methods that enable more precise control over porosity, morphology, and surface chemistry to enhance ion transport and improve the mechanical properties of polymers. Therefore, the chapter thoroughly discusses performance aspects such as electrolyte compatibility, ion selectivity, operational durability, and thermal stability, which are crucial for developing next-generation supercapacitor membrane designs.

## 6.2 MEMBRANE TYPES AND MATERIALS

### 6.2.1 Polyolefin Membranes

Polyolefin-based membranes, particularly PP and PE, are the most common materials used in commercial electrochemical devices such as lithium-ion batteries and supercapacitors. The prevalence of these materials in commercial usage is because of their relatively low cost and wide range of properties, which include chemical

DOI: 10.1201/9781003679363-6

stability against different reagents, mechanical strength, and high resistance to thermal shrinkage and distortion. In general, these membranes consist of a polymer that has been partially and perpendicularly drawn, with micropores that facilitate electrolyte flow but are small enough to prevent direct contact between electrodes. However, the inherent hydrophobicity of polyolefin membranes leads to poor wettability, which negatively impacts electrolyte fluidity and reduces ion uptake. This limitation becomes more pronounced when ionic liquid electrolytes or gel polymer electrolytes are considered for next-generation supercapacitors. Since electrolyte penetration into the pores is hindered by this poor wettability, the contact area between the electrolyte and the electrode is reduced, which in turn increases the equivalent series resistance (ESR) and leads to a decrease in capacitance. To overcome this drawback, various surface modifications such as plasma treatment, ceramic coating, and polymer grafting are usually adopted to increase the affinity and compatibility of the membrane surface with the electrolyte solution. These modifications involve various chemical treatments that result in the formation of strong interfacial interactions between the hydrophilic polymer chains and the membrane surface. Recently, much attention has been given to the development of multi-layered polyolefin membranes that consist of PE and PP layers. This membrane possesses the advantage of a thermal shutdown feature, where the pores collapse to stop the transport of ions when the temperature is high [4]. For a supercapacitor-based energy storage system, especially where flexible or high-rate operation is common. These multilayer designs are continually being optimized for safety and performance. Despite their ease of production and durability, polyolefin membranes remain a primary material in supercapacitors. Moving forward, their development will increasingly focus on hybridization and functionalization strategies to overcome wetting limitations, making them more compatible with various electrolyte systems [5].

### 6.2.2 PVDF and PVDF-HFP Membranes

PVDF and its copolymer PVDF-HFP have emerged as promising materials for supercapacitor membranes, mainly due to their excellent electrochemical stability, great chemical compatibility with organic electrolytes, and inherent polarity, as shown in Figure 6.1.

Unlike hydrophobic polyolefins, PVDF-based polymers show improved compatibility with liquid electrolytes, hence improved electrolyte retention, ionic conductivity, and reduced interfacial resistance. Moreover, the high porosity of PVDF-based polymers makes them suitable hosts for flexible or solid-state supercapacitors. This is because PVDF-based polymers can function as hosts in gel polymer electrolytes. For instance, PVDF-HFP has been shown to benefit from the incorporation of hexafluoropropylene units. This has been attributed to the fact that the incorporation of these units results in an amorphous structure that is softer in terms of mechanical properties [7]. As a result, PVDF-HFP has been shown to exhibit improved electrolyte retention and ionic conductivity. This has made it the most commonly used polymer in gel-based supercapacitors. However, it has been noted that PVDF-based membranes exhibit poor mechanical properties. Their softer nature and lower tensile properties, compared to denser polyolefin membranes, limit their use in high-pressure or high-temperature

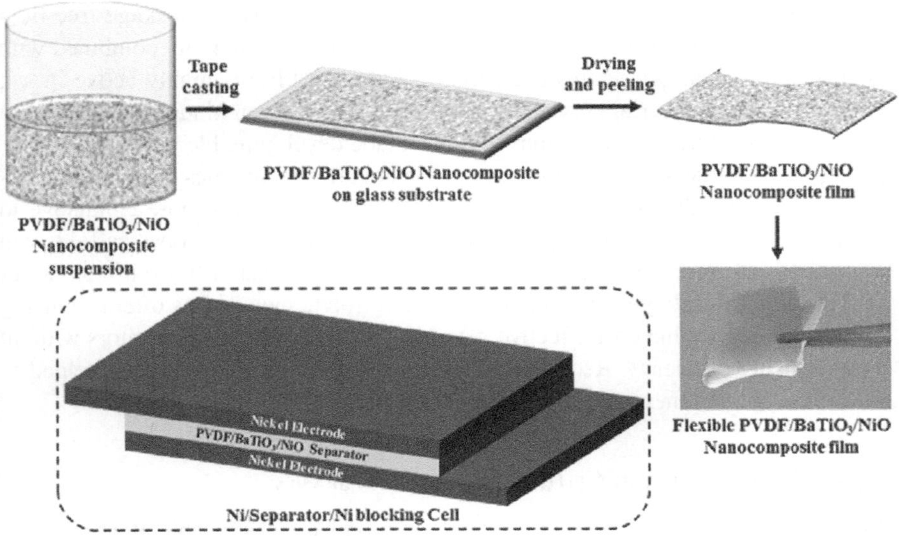


**FIGURE 6.1** Schematic representation of membrane fabrication.

**(Reproduced with permission [6].)**

conditions unless reinforced. To address these shortcomings, composite strategies have become the predominant approach. This usually includes the integration of ceramic nanoparticles like $SiO_2$, $Al_2O_3$, and $TiO_2$, carbon-based nanomaterials such as CNTs and graphene, as well as polymer blends like PANs and PEOs. These will provide the necessary mechanical robustness, facilitate the formation of effective ionic transport networks, and minimize the occurrence of leakage problems [8]. Future developments will likely concentrate on the fabrication of hybrid membranes that transcend the traditional balance between conductivity and mechanical robustness.

### 6.2.3 Gel and Solid Polymer Electrolyte Membranes

Gel polymer electrolytes and solid polymer electrolytes have advanced significantly in developing supercapacitor membranes, as they combine the desirable properties of liquid electrolytes and solid supports. The formation of a crosslinked polymer network, which provides continuous pathways for ion transport. For instance, PEO is the most extensively studied polymer for membrane production due to its ability to dissolve salts and ionic liquids, facilitating their coordination through its ether oxygen. However, it should be noted that PEO can only be effectively used when blended with ionic liquids and supported on microporous materials like polypropylene membranes. These PEO-based membranes not only enhance ionic conductivity but also improve interfacial and dimensional stability[9]. GPEs usually comprise a polymer matrix that is fully swollen with a liquid electrolyte. This combination provides the advantage of mechanical flexibility along with high ionic conductivity. In the case of SPEs, the absence of liquid components is compensated for by the segmental motion of the polymer chains and the interactions between the polymer chains and the salts. This advantage is particularly desirable for leakage-free flexible supercapacitors. Nonetheless, achieving high ionic conductivity combined with mechanical robustness remains challenging. Although PEO is highly active in salt solvation processes, certain crystalline structures it forms can hinder ion movement at room temperature [10]. Solutions to this include developing blended electrolytes, adding ceramic fillers, and employing copolymerization techniques that modify crystallinity, increase the amorphous content, and create pathways more conducive to ion migration. Moreover, incorporation of SPEs within microporous scaffolds can be used to improve structural integrity and dimensional stability even after long-term cycling. Overall, gel and solid polymer electrolyte membranes offer promising avenues for developing cost-effective, flexible, and efficient supercapacitors without compromising efficiency. Recent developments in polymer science have helped to move these membranes toward practical applications [11].

### 6.2.4 Biopolymer and Cellulose-Based Membranes

Biopolymer-based membranes are considered promising sustainable alternatives to synthetic polymer membranes for supercapacitors. Specifically, bacterial cellulose, paper, and naturally occurring biostructures such as eggshell membranes combine environmental compatibility with useful functional properties. These are excellent ideas since biopolymers are inherently renewable, recyclable, and cost-effective, aligning well with the growing demand for clean energy storage solutions. Cellulose and its derivatives are especially appealing owing to their numerous surface hydroxyl groups, which promote strong interactions with liquid electrolytes, leading to high wettability [12]. This ensures easy electrolyte uptake and facilitates rapid ionic transport through the membrane. Additionally, cellulose nanofibers can be designed to be highly porous, with interconnected nanofibers forming a lightweight yet mechanically flexible structure. This structure is capable of meeting high ionic conductivity as well as cycling stability. Recent advances in engineering cellulose nanofibers have proved the efficiency of one or multiple layers of membranes, which are highly porous, electrolyte-affinitive, and mechanically robust. These membranes now rival, and in some cases outperform, traditional polyolefin membranes in ionic transport and compatibility with aqueous and organic electrolytes. Current research in cellulose and other biopolymers like chitosan, alginate, and starch-based composites mainly involves the incorporation of functional groups and blends of synthetic polymers for improved conductivity and stability. Biopolymer and cellulose-based membranes have tremendous potential for future high-performance and green supercapacitors. Their key advantage for next-generation energy storage devices, especially where environmental impact and biodegradability are concerns, is their ability to merge green chemistry principles with advanced electrochemical functionalities.

### 6.2.5 Nanocomposite and Nanofiber Membranes

Nanocomposite and nanofiber membranes are among the most advanced types of supercapacitor membranes, providing the ability to design not only ion transport pathways but also mechanical reinforcements at the nanoscale (Table 6.1). The main

**TABLE 6.1**
**Comparison of Different Supercapacitor Membrane Types**

| Membrane Type | Representative Materials | Ionic Conductivity (S $cm^{-1}$) | Key Advantages | Limitations |
|---|---|---|---|---|
| Polyolefin (PP/PE) | PP, PE, PP–PE multilayers | $10^{-5}$–$10^{-4}$ | Low cost, safety (shutdown) | Poor wettability |
| PVDF-based | PVDF, PVDF-HFP | $10^{-4}$–$10^{-3}$ | High polarity, flexibility | Low tensile strength |
| Gel polymer | PEO, PMMA, PAN | $10^{-3}$–$10^{-2}$ | High ionic transport, flexible | Risk of leakage |
| Solid polymer | PEO–Li, PAA-based | $10^{-5}$–$10^{-3}$ | Leakage-free, safe | Low room-temp conductivity |
| Biopolymer | Cellulose, Chitosan | $10^{-3}$–$10^{-2}$ | Renewable, eco-friendly | Moisture sensitivity |
| Nanocomposite | PVDF-HFP/GO, PAN–CNT | $10^{-3}$–$10^{-2}$ | High conductivity & durability | Complex fabrication |

features of electrospun nanofiber structures include a large surface area, interconnected porosity, and adjustable fibre orientation, each of which individually and collectively enhances electrolyte uptake and enables rapid ion mobility. Due to these qualities, electrospun membranes are widely used in both flexible and high-power supercapacitor systems. It has been demonstrated that ionic conductivity, mechanical strength, and electrochemical stability can be enhanced when functional nanomaterials like graphene oxide, CNTs, ceramic nanoparticles, and heteroatom-doped carbons are embedded in polymer matrixes like PVDF-HFP. For instance, GO provides oxygen-containing functional groups, increases the membrane's hydrophilicity, and promotes stronger electrolyte–membrane interactions [13]. However, achieving a uniform distribution of nanofiller is essential, as aggregates can clog the membrane layer and decrease ion mobility and long-term cycling performance. Besides simple additive techniques, hybrid nanofiber membranes have also been designed using co-electrospinning techniques, combining polymers and nanomaterials. The techniques not only improve on past limitations but also allow for fast ion transport, as well as durability against deformations caused by repeated cycles. The nanofibers can be arranged in a way that affects infiltration as well as transport anisotropy, introducing another dimension of flexibility. Nanocomposite and nanofiber membranes can be considered a new generation of supercapacitor membranes, combining desirable physical attributes with superior functionalities. Future research should focus on strengthening filler–matrix interactions, gaining better control over fibre architectures, and refining the electrospinning process for practical device implementation.

## 6.3 ROLES AND KEY CHARACTERISTICS OF THE MEMBRANE

The membrane of the supercapacitor is much more than simply a Separator. It is the determining factor in the device's capacity, safety, and operational life. Its main function is that of separating the positive and negative electrodes. It must be able to prevent the possibility of a short circuit. At the same time, the membrane must allow the free flow of ions from the electrolyte between the electrodes during charge and discharge. This is the ideal balance of properties. It is the determining factor in the ionic flow and distribution. A good membrane must have high ionic mobility. This is the key to the high rate of charge and discharge that characterizes the supercapacitor compared to the battery. However, when ion transport is restricted, the ESR increases, the locally accessible capacitance decreases, and overall power density is reduced.

Moreover, the membrane is the primary electrolyte reservoir for gel and solid-state supercapacitors. Membrane properties, including porosity, wettability, and integrity, play a critical role in the ability of the membrane to absorb and retain the electrolyte. These properties underscore the importance of the membrane in the efficiency, safety, and overall reliability of the supercapacitor. For this reason, high-performance membranes can be tailored for specific functions, including selectivity, mechanical strength, and thermal stability. The membrane is, therefore, not just a barrier for physical separation. Rather, it is a critical part of the supercapacitor for the efficient movement of ions, delivery of power, and overall reliability of the supercapacitor. Thus, the membrane in the next-generation supercapacitors enhances the performance of the devices, and therefore, it is one of the most critical features of the supercapacitors. The characteristics of the supercapacitor membrane, which influence ion transport, device stability, and safety, play a crucial role in the supercapacitor's performance. Membrane effectiveness can be assessed using several interconnected metrics.

### 6.3.1 Ionic Conductivity

High ionic conductivity is essential for fast ion transport and minimizing internal resistance [14]. A good example of solid polymer-ionic liquid composites supported on microporous structures delivers very high ion transport performance, enabling fast charge–discharge cycles and low ESR values.

### 6.3.2 Porosity and Wettability

Porosity and electrolyte affinity are key membrane properties that affect the performance of supercapacitors, as these properties affect the amount of ions available, the path that the ions take, and the ESR. Porous membranes with high wettability properties help in increasing the amount of absorbed electrolyte, hence reducing the internal resistance of the membrane by allowing the free flow of ions across the surface and in the bulk of the membrane [15]. Biopolymers and cellulose-based membranes have shown excellent electrolyte absorption and retention properties. These materials have shown excellent electrochemical properties when hydrophilic groups are introduced, and the materials are fabricated in the form of interpenetrating fibrous networks, thereby showing excellent interface compatibility [5].

### 6.3.3 Mechanical Strength and Thermal Stability

In addition, the membrane has to withstand the mechanical stresses associated with the assembly of the device and the electrochemical cycling. Polyolefin membranes offer advantages in terms of their high resistance to thermal shrinkage, whereas composite gel systems offer the high stability of one component along with the high tensile strength of the other component [16]. These properties are of major importance for flexible and high-rate supercapacitors, in which the mechanical strength is a measure of the overall reliability of the devices. For example, a low mechanical or thermal stability may result in the deformation of the membrane or the loss of contact with the electrodes, and in extreme cases, may cause a short circuit.

### 6.3.4 Device-level Performance

Furthermore, the selection of the membrane also affects the overall performance of the device in terms of capacitance, energy density, power density, ESR, and cycling stability. For example, a polymer-ionic liquid hybrid membrane supported on a porous substrate was found to exhibit exceptionally high specific capacitance. The characteristics of the membrane in terms of its structural arrangement and physicochemical properties play a vital role in the overall determination of the performance and durability of the supercapacitor devices. In general, the membrane can be classified into solid ion conductors, solid polymer electrolytes, ion exchange membranes, and microporous polymeric membranes that are derived from natural and synthetic polymers, including mineral-based and functionalized membrane systems. The overall efficiency of the membrane is mainly governed by several parameters, such as ionic conductivity, ion exchange capacity, chemical and thermal stability, porosity, uptake of the membrane material with the electrolyte solution, thickness, and uptake of the membrane material with the solution. However, the optimization of these parameters is often a critical issue due to the presence of trade-offs between high ionic transport properties and the overall mechanical and dimensional stabilities [17].

### 6.3.5 Cycling Stability and Leakage

Endurance during cycling is conducted over long durations of time and plays a vital role in determining the reliability of membranes. High-performance membranes, in particular, show superior reliability in terms of retaining capacitance after numerous cycles of charging and discharging, in addition to possessing a low self-discharge rate and leakage of the electrolyte. Notably, membranes based on electrospun nanofibers show superior cycling reliability, as evidenced by only a small rise in leakage current even after long cycling times.

## 6.4 PERFORMANCE LIMITATIONS IN SUPERCAPACITORS

Supercapacitors are extensively used in transportation, industries, military equipment, and consumer electronics owing to their high-power density and long cycle durability [18]. Despite the extensive application of supercapacitors, there are certain limitations that are yet to be overcome, and these issues are closely associated with the membranes and the supercapacitor device itself. The first limitation is that the energy density of supercapacitors is typically low, i.e., <20 Wh/kg, whereas the energy density of batteries is between 30 Wh/kg and 200 Wh/kg. The low energy density of supercapacitors makes the device larger and less compact, thus the need for the development of membranes that can provide higher surface area and higher voltage windows [6]. The low voltage of individual cell connections also makes the separation of membranes essential so that the supercapacitor is not overcharged. Thus, the development of supercapacitors has progressed much faster than the standardization of the membranes, electrolytes, safety, performance testing, and recycling of the supercapacitor device itself. Membranes play a crucial role in the development of the next generation of supercapacitors, particularly flexible, miniaturized, hybrid, and multifunctional ones. The increased requirement for portable electronics makes the development of thin, flexible membranes that are mechanically strong, leak-proof, and suitable for gel or solid electrolytes an essential requirement for the development of safe and compact devices [9]. These membranes also promote micro supercapacitor designs by supporting compact integration while preserving high ionic conductivity and electrochemical stability. In hybrid battery–supercapacitor configurations, membrane optimization can improve ion transport, expand the operating voltage window, and enhance interfacial compatibility between battery-type and capacitive electrodes, thereby helping to reconcile high-power and high-energy density. Additionally, advanced membranes can facilitate new technology trends such as self-healing and transparent energy storage devices by providing a platform for multifunctionality, structural robustness, and mechanical stability after deformation [10]. Therefore, membrane design can be highlighted as an important enabler for future supercapacitors in microelectronics and advanced multifunctional energy storage devices.

## 6.5 FUTURE PERSPECTIVES ON SUPERCAPACITOR MEMBRANE

The concept of designing the membrane for the supercapacitor inherently requires the balancing of several trade-offs for the primary parameters that define the efficiency, reliability, and safety of the supercapacitor [11]. For instance, high ionic conductivity along with low ESR is critical for high rates of charge/discharge cycles, high-power density, and low losses. However, this requires the material to be highly porous. These compromises are evident in the representative fabrication strategies summarized in Table 6.2. These properties are often accompanied by the need for highly porous and interconnected structures to enable the efficient transport of ions. However, high porosity can also limit the membrane's selectivity, mechanical robustness, and dimensional stability. In addition, the membrane must be able to withstand cyclic bending, swelling, and compression for wearable devices, as well as thermal stability for high-power or high-temperature applications. Additionally, the integration of green, biodegradable, or recyclable materials for environmental sustainability introduces further challenges in preserving long-term durability and performance consistency [12].

However, these trade-offs are being addressed by newly developing membrane technologies, including bacterial cellulose membranes, membranes derived from biomass materials, hybrid membranes composed of ionic liquids, polyethylene oxide, and microporous polymers, nano-engineered membranes, additive manufacturing technologies, and smart membranes. Bacterial cellulose membranes, as well as membranes derived from biomass materials, offer environmentally friendly membranes with superior wettability and well-developed porous structure. Hybrid membranes composed of ionic liquids, polyethylene oxide, and microporous polymers exhibit near-liquid ionic mobilities, as well as the mechanical robustness of solid-state materials [13]. Nano-engineered membranes, including those with graphene oxide, MXenes, and ceramic materials, show superior conductivity, as well as well-controlled pore structure. Additive manufacturing techniques, including 3D printing and inkjet printing, play a significant role in developing hierarchical membranes, including those for flexible devices. Notably, smart membranes, including self-healing membranes under mechanical stress and thermal shutdown membranes at high temperatures, offer a multifunctional solution, including efficiency, safety, and lifetime [14]. Overall, these newly developing membranes show a trend moving away from membranes that are merely conductive and robust to membranes that are adaptive, sustainable, and well-suited to next-generation supercapacitor technologies.

## 6.6 CONCLUSION

Membranes in supercapacitors are no longer merely passive separators. They are crucial for safe device operation, long-term durability, and efficient ion transport. The diversity of materials used varies from relatively inexpensive polyolefins and PVDF-based materials that are versatile and easy to modify to advanced materials such as nanocomposites and biopolymer-derived materials, with an emphasis on sustainability. This is an indication that there is a need to develop membranes that are tailored to specific needs in their applications, either in flexible and lightweight energy sources or in high-energy and long-lasting storage devices. The recent developments in membrane fabrication processes, such as electrospinning, bilayer structures, composite coatings, and additive manufacturing, have shown the potential of such processes in creating specific membrane properties such as porosity and functionality. Future supercapacitor technologies will be driven by the use of eco-friendly materials, nanostructured designs, and smart membranes, enabling higher capacity, better safety, broader functionality, and improved sustainability.

**TABLE 6.2**
**Representative Fabrication Methods and Their Features**

| Fabrication Method | Key Materials | Structural Characteristics | Advantages | Limitations |
|---|---|---|---|---|
| Electrospinning | PVDF-HFP/GO | Nanofibrous, high porosity | High ion mobility | Filler agglomeration |
| Bilayer Formation | Cellulose/PVA | Double-layered | Improved wettability | Complex prep |
| Solid Electrolyte Infiltration | PEO/Li-salt | In-situ filled pores | Zero leakage | Slow diffusion |
| Composite Coating | Eggshell–$TiO_2$ | Smooth functionalized surface | Better interface control | Thin coating limits ions |

## REFERENCES


[1] F. Cheng, C. Wu, S. Wang, and S. Wen, "Polydopamine-modified electrospun polyvinylidene fluoride nanofiber based flexible polymer gel electrolyte for highly stable dye-sensitized solar cells," *ACS Omega*, vol. 6, no. 43, pp. 28663–28670, 2021, doi: 10.1021/acsomega.1c03232

[2] I. O. Riwa, A. T. Banyikwa, R. Costa, A. Veved, N. S. Babu, and M. G. Sahini, "Advances in copolymers based on polyvinylidene fluoride (PVDF) and their composites for piezoelectric energy harvesting," *Resour. Chem. Mater.*, p. 100124, 2025, doi: 10.1016/j.recm.2025.100124

[3] C. W. Nan, L. Fan, Y. Lin, and Q. Cai, "Enhanced ionic conductivity of polymer electrolytes containing nanocomposite [formula presented] particles," *Phys. Rev. Lett.*, vol. 91, no. 26, pp. 1–4, 2003, doi: 10.1103/PhysRevLett.91.266104

[4] Z. Liu et al., "Safer lithium-ion batteries from the separator aspect: Development and future perspectives," *Energy Environ. Mater.*, vol. 4, no. 3, pp. 336–362, 2021, doi: 10.1002/eem2.12129

[5] S. Ahankari, D. Lasrado, and R. Subramaniam, "Advances in materials and fabrication of separators in supercapacitors," *Mater. Adv.*, vol. 3, no. 3, pp. 1472–1496, 2022, doi: 10.1039/d1ma00599e

[6] M. Sharma, and A. Gaur. "Fabrication of PVDF/BaTiO3/NiO nanocomposite film as a separator for supercapacitors," *J. Energy Storage.*, vol. 38, p. 102500, 2021, doi: 10.1016/j.est.2024.113781

[7] H. Rashid Khan and A. Latif Ahmad, "Supercapacitors: Overcoming current limitations and charting the course for next-generation energy storage," *J. Ind. Eng. Chem.*, vol. 141, no. May 2024, pp. 46–66, 2025, doi: 10.1016/j.jiec.2024.07.014

[8] N. P. D. Ngidi, A. F. Koekemoer, and S. S. Ndlela, "Recent advancement in the electrochemical performance of electrochemical capacitors based on biomass-derived porous carbon: A review," *J. Energy Storage*, vol. 89, no. March, p. 111638, 2024, doi: 10.1016/j.est.2024.111638

[9] X. Li et al., "Poly (ethylene oxide) based solid polymer electrolyte improved by multifunctional additives of poly (acrylamide) and LiI," *Electrochim. Acta*, vol. 445, no. December 2022, p. 142062, 2023, doi: 10.1016/j.electacta.2023.142062

[10] A. Alvarez-Fernandez, G. Hernández, and J. Maiz, "Unlocking solid polymer electrolytes: Advancing materials through characterization-driven insights," *JACS Au*, 2025, doi: 10.1021/jacsau.5c00442

[11] T. K. Nguyen, "Pham-Truong TN. Recent advancements in gel polymer electrolytes for flexible energy storage applications," *Polymers.*, vol. 16, no. 17, p. 2506, 2024, doi: https://doi.org/10.3390/polym16172506

[12] E. Lizundia, C. M. Costa, R. Alves, and S. Lanceros-Méndez, "Cellulose and its derivatives for lithium ion battery separators: A review on the processing methods and properties," *Carbohydr. Polym. Technol. Appl.*, vol. 1, no. May, p. 100001, 2020, doi: 10.1016/j.carpta.2020.100001

[13] A. Alhinai, M. Edokali, L. Tliba, R. Menzel, and A. Hassanpour, "Performance comparison of differently dried graphene oxide-based membranes for desalination by forward osmosis," *Chem. Eng. Sci.*, vol. 317, no. December 2024, p. 121998, 2025, doi: 10.1016/j.ces.2025.121998

[14] F. Zhao et al., "Internal resistance reduction strategies for high-power and fast-charging Lithium-ion batteries," *Mater. Sci. Eng. R Reports*, vol. 166, no. August, p. 101076, 2025, doi: 10.1016/j.mser.2025.101076

[15] V. Gregorio, J. H. Chang, N. García, and P. Tiemblo, "Porous membranes of polymer blends and their use as functional separators for electrochemical batteries," *J. Memb. Sci.*, vol. 730, no. April, p. 124187, 2025, doi: 10.1016/j.memsci.2025.124187

[16] A. Bin Rashid, M. Haque, S. M. M. Islam, and K. M. R. Uddin Labib, "Nanotechnology-enhanced fiber-reinforced polymer composites: Recent advancements on processing techniques and applications," *Heliyon*, vol. 10, no. 2, p. e24692, 2024, doi: 10.1016/j.heliyon.2024.e24692

[17] A. L. Ahmad, U. R. Farooqui, and N. A. Hamid, "Effect of graphene oxide (GO) on Poly(vinylidene fluoride-hexafluoropropylene) (PVDF- HFP) polymer electrolyte membrane," *Polymer (Guildf).*, vol. 142, pp. 330–336, 2018, doi: 10.1016/j.polymer.2018.03.052

[18] L. Zhang, Y. Wang, and C. Chang, "Effects of freezing/thawing cycles and cellulose nanowhiskers on structure and properties of biocompatible starch/pva sponges," *Macromol. Mater. Eng.*, vol. 295, no. 2, pp. 137–145, 2010, doi: 10.1002/mame.200900212

# 7 Membranes for Batteries

## 7.1 INTRODUCTION

Membranes play an important role in governing the performance, safety, and longevity of modern battery systems when they are integrated within solid or liquid electrolytes [1]. The primary role of membranes is to physically isolate electrodes while allowing selective ion transport, and it also facilitates ionic conductivity, thermal shrinkage resistance, electrolyte wettability, and electrochemical stability. Most of the current research focuses on the following batteries: Lithium-ion batteries (LIBs), sodium-ion batteries (SIBs), and solid-state batteries (SSBs). LIBs membranes provide high ionic conductivity and prevent dendritic short circuit by acting as a physical barrier [2]. During the charging process, lithium ions are extracted from the cathode materials and intercalated into the anode (usually graphite), whereas in the discharging process, they migrate back and reinsert into the cathode materials. This reversible shuttling of $Li^+$ ions between these electrodes reinforces the charge–discharge process, which allows electrochemical energy storage and the conversion between chemical and electrical energy in LIBs. The electrolyte in a lithium-ion battery serves as both an ionic conductor and an electronic separator between the cathode and anode. Further challenges in SIBs membranes due to larger ion sizes and solvent compatibility, and for SSBs will require materials that act as both electrolytes and membranes, maintaining prominent ionic conductivity with mechanically and thermally stable materials in extreme conditions [3]. This chapter explains the utilization of membranes in various rechargeable batteries, focusing on their structural frameworks and materials used, functional demands, and fabrication methods. It discusses performance differences among LIBs, SIBs, and SSB and also introduces new concepts such as polymer–ceramic composites, functional coatings, and solid polymer electrolytes. Additionally, the chapter emphasizes the interaction between membrane architecture, mechanisms of ion transport, and ionic stability under high-voltage and high-temperature conditions.

## 7.2 MEMBRANE TYPES AND MATERIALS IN LIBS

LIBs are the electrochemical cells that simultaneous conduct electrons through an external circuit and cyclic migration of lithium-ion ($Li^+$) ions in the electrolyte, across a porous separator (membrane) in order to balance the electrochemical reaction between the active materials adhered to each electrode. The primary function of the membrane is to physically separate the cathode and anode by preventing direct electrical contact and short circuits. It also plays an important role as a multifunctional component in safety, ion transport, and long-term durability [4, 5]. The components of the batteries include a separator, a solid ion-conducting membrane, a solvent, an electrolyte, and other additives. An electrolyte should be stable in highly oxidative and reductive environments, highly conductive to ions, and capable of supporting


DOI: 10.1201/9781003679363-7

electrochemical reactions. In liquid-electrolyte systems, a porous membrane prevents direct contact between the electrodes while allowing ion transport. Alternatively, the membrane itself can be a $Li^+$ conductor, eliminating the need for a separate liquid electrolyte, or the liquid electrolyte can be incorporated into a polymer matrix to form a gel. These membranes play a critical role in the secure separation of the electrodes while facilitating efficient ion movement in a lithium-ion battery. The membrane's pore diameter and tortuosity significantly influence the battery's ionic conductivity and internal resistance. Secondly, the polyolefin membrane, which is normally either PP or PE, has the property of "thermal shutdown," which means that when the temperature is increased to about 130–140°C, the pores tend to shrink due to the melting of the polymer, which stops the movement of $Li^+$ ions and protects the battery from overheating [6–8]. The key factor in determining the materials for membranes in LIBs is cost, mechanical properties, ionic transport properties, and safety properties. Even though much emphasis is placed on polymer materials, composite materials, and green materials, microporous polyolefin materials are still dominant in the market for LIBs. The leading technology for polymer membranes in LIBs is represented by microporous polyolefin membranes, which include PE and PP membranes.

### 7.2.1 Microporous Polyolefins

These materials have been commonly used for membranes in LIBs at large scales due to their resistance to different kinds of degradation, stability in different kinds of electrolyte systems, and their ability to be produced by extrusion and stretching techniques. Due to the semicrystalline nature of the materials, they show characteristics such as high tensile strength and micropores, which are favourable for $Li^+$ conduction. However, the hydrophobic nature of the materials is a disadvantage since they are not easily wetted by the electrolytes. Moreover, there is the problem of dimensional shrinkage at high temperatures, which is a major safety hazard in electric vehicles.

### 7.2.2 Liquid-Electrolyte Systems

Generally, in traditional SIBs with liquid electrolytes, membranes are used from the LIB domain and consist mainly of microporous polymeric membranes such as polyolefin membranes from polyethene (PE) or polypropylene (PP). These membranes are of prime importance because of their mechanical properties and low cost. However, because of the poor diffusivity and solvation energies associated with $Na^+$ ions, these membranes need to be optimized for $Na^+$ ion conduction [9]. Surface modification techniques, such as the application of a ceramic coating or the grafting of polar polymers, may improve the wetting characteristics and the contact resistance, thus facilitating the transportation of the sodium ions.

### 7.2.3 Fluorinated Polymers

The PVDF and its copolymer with hexafluoropropylene (PVDF-HFP) exhibit enhanced wettability and polarity compared to the polyolefin materials, which will help in the fast transportation of ions. These materials are usually employed in multilayer configurations, where they are laminated with cores of PE and PP for enhanced thermal safety and electrolyte wettability. However, their high cost restricts their application and prevents their replacement with conventional materials.

### 7.2.4 Modified and Composite Membranes

To overcome these intrinsic limitations of hydrophobic polyolefin separators, ceramic surface modifications are commonly applied, which are generally performed by applying oxide coatings such as $SiO_2$, $Al_2O_3$, or $TiO_2$, which are dominant in the market at present. As illustrated in Figure 7.1(a), a dual-coating strategy can be applied to a polyethylene (PE) membrane, where an $Al_2O_3$ layer deposited by atomic layer deposition (ALD) is followed by a polydopamine (PDA) coating, forming a polymer–ceramic composite structure. These coatings have the advantage of raising the surface energy of the membrane, thus raising the wettability of the electrolyte and its thermal and mechanical stability. The resulting microstructure is evidenced by the top-view and cross-sectional SEM images in Figure 7.1(b), which confirm the preservation of a porous architecture after coating. The ceramic coatings not only

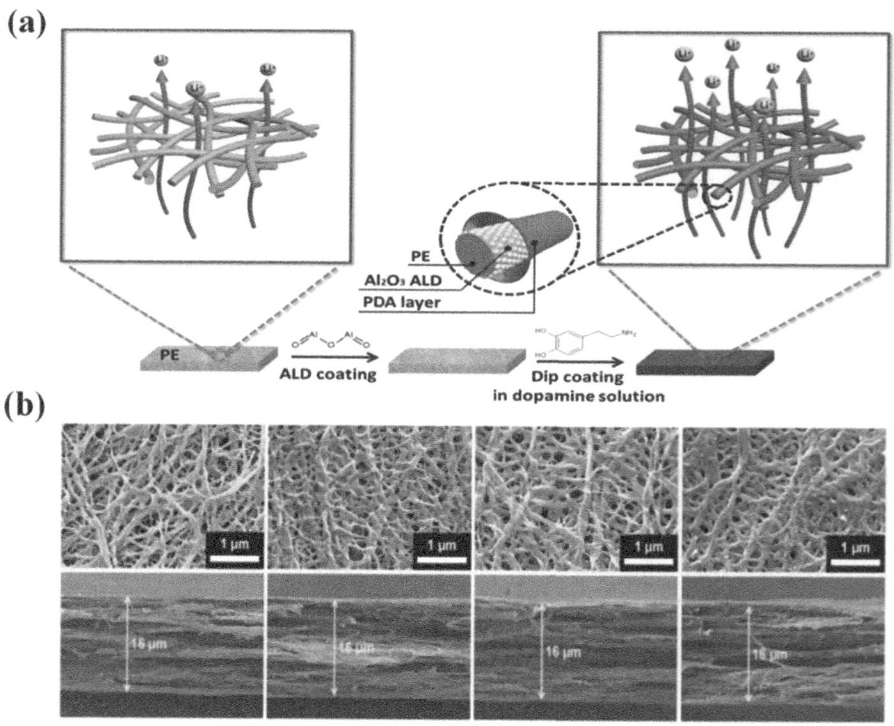


**FIGURE 7.1** Schematic illustration of the preparation of the composite membranes and SEM images and corresponding cross-sectional views of composite membranes. (a) Schematic representation of the dual-coating process. (b) Top-view and cross-sectional SEM images showing the porous structure of the membrane after coating.

**(Reproduce with permission [10].)**

improve interfacial affinity with the electrolyte but also offer a blocking effect to the penetration of lithium dendrites, thus extending the service life of high-density power sources. More broadly, hybrid polymer–ceramic composite membranes incorporating inorganic nanofillers are considered a promising route to achieving an optimal balance between flexibility and dimensional stability.

### 7.2.5 Bio-Based Membranes

The demand for sustainable materials has boosted the adoption of cellulose-based membranes. These membranes exhibit strong hydrophilicity for fast electrolyte uptake, and their engineered nanofibrillar architectures can attain porosities of up to ~70%, thereby minimizing ion diffusion distances.[11]. Besides, due to the hydrogen bonding in the networks, these bio-based membranes have high thermal stability and mechanical strength. Although several challenges remain regarding large-scale production and compatibility with non-aqueous electrolytes, bio-based membranes represent a promising approach for developing more environmentally sustainable lithium-ion battery membranes.

## 7.3 SODIUM-ION BATTERIES MEMBRANE

The widespread availability and even geographical distribution of sodium resources have made sodium-ion batteries (SIBs) attractive candidates as cost-effective and sustainable alternatives to LIBs [12]. However, sodium ions have a larger ionic radius ($Na^+$ ~1.02 Å vs. Li 0.76 Åfor $Li^+$), which makes membrane and electrolyte design even more complicated and strongly influences ionic transport efficiency, structural integrity, and long-term cycling stability [13]. The Progress in the SIB membrane technology is occurring in two ways, including the development of polyolefin membranes for liquid-electrolyte systems, and the fabrication of solid ceramic or composite membranes for solid-state designs [14]. Both methods address the common problems, which are the need for an appropriate approach to the problems posed by the transport of $Na^+$ at the interface, as well as the mechanical durability of the membranes.

### 7.3.1 Solid-State Sodium-Ion Batteries (SSIBs)

A significant step change in the field of energy storage technology, the move from liquid electrolyte-based cells to solid-state cells, provides enhanced safety benefits, increased energy density, and the benefits of more stable materials. The solid-state cell employs a solid electrolyte membrane as the ionic conductor and the electric separator. This places very strict demands on the selection of materials. Inorganic ceramic electrolytes such as sulfides and garnet-type oxides have shown ionic conductivity > $10^{-3}$ S/cm, rivalling the performance of liquid electrolytes [15]. However, these materials' brittleness and poor contact with the electrodes make them unsuitable for practical application. Polymer electrolytes, such as PEO-type or PVDF-type electrolytes, offer good mechanical properties along with good contact with the electrodes but suffer from poor ionic conductivity at room temperature [16]. Hybrid composite materials have been reported for achieving a suitable balance in mechanical properties and ionic conductivity. In this regard, solid-state sodium-ion batteries

(SSIBs) heavily rely on the electrolyte membrane for ionic conductivity, mechanical properties, and stability at interfaces. As SIBs transition from a conventional aqueous or liquid electrolyte-based battery to a solid-state battery, $Na^+$-conducting solid electrolytes have emerged as a key research area and can be categorized as ceramic-type electrolytes, polymer-type electrolytes, and composite or hybrid electrolytes [17].

### 7.3.2 Electrolyte Classification

Three classes of SSB materials can be identified, namely: inorganic ceramics, polymer electrolytes, and composite systems, each with its own set of advantages and disadvantages, thereby influencing the scope of their applications.

### 7.3.3 Inorganic Ceramics

The ionic conductivities of the ceramic electrolyte materials, especially the garnet-type oxides such as LLZO, NASICON-type frameworks, and glasses based on sulfides, are in the range of $10^{-3}$ S/cm, which puts these materials in stiff competition with the conventional liquid electrolytes. These materials show good thermal stability and electrochemical windows, which enable them to show good compatibility with the high-voltage type cathodes. Although the natural brittleness and grain boundary effects, as well as the lack of close contact with the electrodes, still dominate the usage of these materials on a larger scale. In addition, ceramic interfaces can be highly prone to problems such as void formation and dendrite penetration if not designed appropriately [18].

### 7.3.4 Polymer Electrolytes

Polymer-based electrolytes, such as PEO complexed with lithium salts and the PVDF-based systems, are beautiful things to behold in terms of the ease and flexibility in the process and the ability to achieve a conformal contact with the electrodes. They are not only lightweight but can also be easily transformed into thin films and follow many strategies commonly employed in the fabrication of conventional porous membranes. The main drawback in the use of polymer-based electrolytes is the low ionic conductivity at room temperature. Most often, in order to efficiently use polymer-based electrolytes, they are required to be subjected to a high-temperature regime (>60°C). Moreover, the mechanical softness of the polymer-based electrolytes makes them a soft target for dendrite penetration at high current densities [19].

### 7.3.5 Composite Electrolytes

Hybrid electrolytes have been recognized as promising substitutes for the conventional single-phase system, as they combine the mechanical flexibility of the polymer with the mechanical rigidity of the ceramic. Hybrid materials have been designed to exploit the cooperative interaction of the flexible polymer and the rigid ceramic particles, thus providing a system with excellent mechanical strength and ionic conductivity. Hybrid materials consisting of ceramics like LLZO, $Al_2O_3$, and $TiO_2$ in polymers like PEO and PVDF have the advantages of reducing the crystallinity of the polymer, increasing the ionic conductivity, and providing dimensional stability at working temperatures. Besides reinforcing the mechanical core of the system, these particles are also capable of reducing interfacial resistance through the accommodation of the volume changes of the electrodes in the charge and discharge cycles. Due to the presence of a suitable balance of high ionic conductivity, mechanical integrity, and processability, composite polymer–ceramic electrolytes are considered one of the most promising approaches for the development of safe and efficient batteries soon [20, 21].

## 7.4 ROLES AND KEY CHARACTERISTICS OF THE MEMBRANE

Membranes, also known as separators, are important battery parts that are placed between the anode and cathode of a battery to ensure that there is no direct electrical contact between the anode and cathode while at the same time ensuring the passage of ionic charge carriers required to facilitate electrochemical reactions. The membranes, composed of microporous polymers, are important constituents of liquid-electrolyte batteries as well as jelly-roll batteries [22]. The membranes' properties, both physical and physicochemical, are known to affect important battery performance characteristics such as power density, energy density, cycle life, stability, safety, and battery operation. The wettability of the electrolyte makes it easier for the distribution of the electrolyte to be even on the electrodes while at the same time ensuring that the low resistance is maintained; however, the lack of uptake of the electrolyte makes the movement of the ions difficult, leading to polarization and decreased power density. In an attempt to resolve the problems associated with the battery, various surface treatments such as the use of ceramics and polar polymers have been proposed in order to enhance the battery's interaction with the electrolyte and ensure the reliable conduction of $Li^+$ ions. The lithium battery membrane needs to be optimal in terms of performance, security, and strength through the working conditions as a result of the many parameters that compete with each other (Table 7.1).

**TABLE 7.1**
**Key Challenges and Research Strategies in Solid-state Electrolytes**

| Challenge | Cause | Consequence | Mitigation Strategy |
|---|---|---|---|
| **High Interfacial Resistance** | Poor contact between electrode & solid electrolyte | Reduced power density | Apply interlayers, soft polymer coatings |
| **Dendrite Penetration** | Local current hotspots | Short circuit, safety hazard | Uniform $Li^+$ flux, ceramic fillers, ALD coatings |
| **Mechanical Fragility** | Brittle nature of ceramics | Cracking & delamination | Composite or flexible polymer frameworks |
| **Low Room-Temperature Conductivity** | Crystalline polymer domains | Poor ion mobility | Use amorphous polymer blends, ionic liquids |
| **Scalability & Cost** | Complex synthesis routes | Limited industrial adoption | Solution casting, tape casting, 3D printing |
| **Moisture Sensitivity** | Hydrolysis reaction | HF gas, degradation | Encapsulation, inert-atmosphere processing |

### 7.4.1 Thickness

The thickness of commercial membranes is generally between 15 and 25 μm. The thickness can be reduced, which in turn causes the equivalent series resistance (ESR) to be lowered, and power density is increased because the $Li^+$ migration path is shortened. However, excessively thin membranes are prone to mechanical failure, such as tearing during assembly or the formation of pinholes, which can compromise safety. Consequently, optimization of thickness is application dependent.

### 7.4.2 Porosity and Pore Size Distribution

High porosity is essential in lithium-ion battery materials, as it facilitates the wetting of the materials by the electrolyte, as well as low-resistance channels of $Li^+$ transport. At the same time, pores must be narrow enough to prevent penetration of lithium dendrites, hence minimizing the possibility of internal shorts. However, the uniformity of the membrane in terms of the distribution of pore size and tortuosity mainly depends on the morphology of the pores. Although porosity plays a significant role in the efficient transport of ions, it sometimes leads to reduced mechanical strength.

### 7.4.3 Thermal Shrinkage and Mechanical Strength

Dimensional and structural integrity of membranes, particularly under elevated temperatures or mechanical stress, is essential because it is directly linked to the overall safety of the battery. As summarized in Table 7.2, properties such as thickness, porosity, pore size, mechanical strength, and thermal shrinkage collectively govern the balance between ionic transport and mechanical reliability. Although polyolefin membranes are widely employed due to their excellent mechanical properties and

**TABLE 7.2**
**Comparison of Key Requirements and Properties of Battery Membranes**

| Parameter | Lithium-Ion Batteries | Sodium-Ion Batteries | Solid-State Batteries |
|---|---|---|---|
| **Primary Role** | Membrane between electrodes, enabling $Li^+$ transport | Similar function but with $Na^+$ ions | Acts as both a membrane and a solid electrolyte |
| **Ionic Species** | $Li^+$ (~0.76 Å) | $Na^+$ (~1.02 Å) | $Li^+$, $Na^+$, or other ions (solid-state conduction) |
| **Ionic Conductivity** | $10^{-3}$–$10^{-4}$ S $cm^{-1}$ (liquid electrolyte-soaked) | $10^{-4}$–$10^{-5}$ S $cm^{-1}$ | $10^{-3}$ S $cm^{-1}$ (ceramics), $10^{-4}$ S $cm^{-1}$ (polymers) |
| **Operating Temperature** | −20–60°C | −20–80°C | Up to 150°C |
| **Thermal Stability** | Moderate (melts at 130–140°C for PP/PE) | Moderate | Excellent (no melt phase) |

*(Continued)*

**TABLE 7.2**
**(Continued)**

| Parameter | Lithium-Ion Batteries | Sodium-Ion Batteries | Solid-State Batteries |
|---|---|---|---|
| **Mechanical Strength** | High (15–25 μm thick) | Moderate to high | Variable (depends on polymer or ceramic content) |
| **Electrolyte Wettability** | Poor for polyolefins; improved with coatings | Moderate | Excellent for polymer composites |
| **Major Challenges** | Shrinkage, dendrite formation | $Na^+$ solvation & sluggish diffusion | Brittleness, high interfacial resistance |
| **Emerging Solutions** | Coatings, composite designs | Polymer–ceramic hybrid electrolytes | Interface engineering, thin-film designs |

processability, they tend to experience significant thermal shrinkage when the temperature exceeds approximately 120°C, which can potentially lead to internal short circuits within the cell. To address this limitation, strategies such as the application of composite ceramic coatings or the construction of multilayer architectures have been developed [23]. These strategies are highly effective in mitigating the issue of thermal deformation while concurrently increasing the tensile strength and puncture resistance of the membrane itself. The concern of mechanical robustness becomes even more critical for membranes that must endure the mechanical stresses arising from electrode swelling and repeated charge–discharge cycling.

## 7.5 PERFORMANCE LIMITATIONS IN BATTERIES

The progress of membranes used in next-generation batteries has always been influenced by the fluctuating balance between challenges and opportunities. A critical challenge in solid electrolytes is their insufficient ionic conductivity to meet the demands of practical operation at room temperature. In this context, recent progress in garnet- and sulfide-based electrolytes has raised ionic conductivities to near-liquid levels, bringing fully solid-state battery performance closer to reality, as summarized in Table 7.3. Another important issue in device integration is the management of interfaces, as high interfacial resistance and penetration of dendrites not only affect its performance but also pose serious safety risks. Mechanical durability, like other key issues, brings the device to a bottleneck situation as ceramic-based electrodes are inherently fragile and susceptible to mechanical failure upon repeated cycling and are bound to break. Therefore, flexible sulfides and polymer–ceramic hybrid materials are actively being researched, as these materials have the potential for providing mechanical durability and inherent structural recoverability. As such, the scalability of the manufacturing process becomes an important factor for the successful commercialization of the technology, considering the problems associated with the simple, cost-effective, and defect-free fabrication of thin layers of electrolyte.

**TABLE 7.3**
**Performance and Optimization Strategies in LIB Membranes**

| Property | Typical Range | Advantages | Disadvantages | Optimization Strategy |
|---|---|---|---|---|
| **Thickness** | 15–25 μm | Ion transport & power density | Mechanical strength | Use multilayer or reinforced films |
| **Porosity** | 35–60% | Ionic conductivity | Tensile strength | Balance via controlled stretching |
| **Pore Size** | 0.03–0.1 μm | Electrolyte permeability | Dendrite resistance | Narrow distribution control |
| **Thermal Shrinkage** | <5% (ideal) | Dimensional stability | – | Ceramic coatings, crosslinking |
| **Mechanical Strength** | >100 MPa | Safety, puncture resistance | Flexibility | Fibre reinforcement, hybrid design |
| **Electrolyte Uptake** | 100–300% | Ionic conductivity | Swelling risk | Polar polymer blending |

The development of the manufacturing infrastructure and the initial deployment of automotive applications are the current markers for the feasibility of the technology for growth. The possibility of factory establishment and the introduction of the first automotive vehicles that facilitate and speed up adoption are the current signs of industrial feasibility and the future progress of the market. Confronting technical difficulties, these trends and innovations show the direction of technological evolution programmed as the going global era of safe, high-performing, and commercially scalable battery technology unified by solid and hybrid electrolyte membranes.

## 7.6 FUTURE PERSPECTIVES ON LIB MEMBRANES

The performance of LIBs is required to rise in the hard way that electric mobility, grid storage, or portable electronic applications are the ones causing this. Consequently, conventional polyolefin membranes are not satisfactory anymore. The advanced membrane strategies that were developed to overcome these barriers dwell on surface functionalization, ion selectivity, dendrite suppression, and responsive coatings.

### 7.6.1 Functionalization and Surface Grafting

Surface treatments enhance the compatibility of polyolefin membranes with the electrolyte because of their anti-hydrophobic behaviour and limited compatibility [18]. Techniques such as plasma treatment, dopamine-inspired coatings, or radiation grafting can significantly improve the electrolytes' wettability and adhesion to electrodes by adding polar functional groups [19]. Polymer or ceramic coatings applied in situ not only improve uniformity but also decrease the risk of delamination during cycling. These methods also allow the use of polymer materials with desired functionalities, such as polyethylene glycol or polyacrylonitrile, which offer high affinity for $Li^+$ ions and promote the efficient transport of other ions.

### 7.6.2 Ion-Selective and Dendrite-Blocking Coatings

The problem of lithium dendrites forming is the most critical one in high-energy LIBs. In the past, advanced membranes would now pull in metal ($Al_2O_3$, $TiO_2$, $SiO_2$), organic frameworks (MOFs), polymers, along with ceramic particles to make a smooth atomic surface [21]. Such additives and fillers not only lower the local current density but also regulate the distribution of the ions at the electrode-membrane interface, resulting in a uniform $Li^+$ flux. For instance, MOF-related coatings offer the advantages of well-defined nanochannels of precisely the size required for selective $Li^+$ transfer while excluding other larger and more heavily solvated species, resulting in the generation of dendrite along with the progression of these stays at a minimum.

### 7.6.3 Thermally Responsive Coatings

To increase the level of security above the basic "shutdown" feature of polyolefins, the usage of thermally responsive layers is becoming more popular. By shrinking or reorganizing as a response to high temperature, these coatings temporarily eliminate ionic conduction in the affected area to avoid a chain reaction, while at the same time, they maintain their mechanical properties throughout normal operation. Those smart polymer systems that can undergo reversible transitions have a much more regulated shutdown effect compared to the certainly non-reversible one, which is obtained through conventional films where pores collapse.

## 7.7 CONCLUSION

The membrane technology in batteries covers a broad spectrum of materials and functional properties, all of which are suited to the specific demands of various electrochemical materials. For lithium-ion batteries (LIBs), microporous polyolefin membranes remain the industry standard, while ceramic or polymer coatings are widely employed to enhance their thermal stability, electrolyte wettability, and safety. The membranes ensure adequate safety and stability, but their performance is restricted by thermal shrinkage and poor wettability. Although sodium-ion batteries (SIBs) are an emerging technology, significant advances have been achieved in separator and solid-state electrolyte design through the incorporation of pyrogenic alumina and carbon-based interlayers to improve $Na^+$ ion transport and interfacial stability. Membrane has been fundamentally redefined for lithium-based all-solid-state batteries (SSBs), where the former polymer membrane is replaced by the latter ceramic or glassy solid electrolytes acting as ionic conductors as well as a physical blockade. However, despite all the advancements in this area, there are still problems in the interface contact and mechanical robustness for this type of material. In fact, as discussed in this chapter, the performance of the membrane is based on complex trade-offs in terms of its safety, transport, and mechanical characteristics. Moving forward in the future, more research in the field of hybrid and post-lithium electrolytes and the development of efficient and effective processing routes are essential in the development of future energy storage devices. Future progress will rely on sustained research into hybrid and post-lithium electrolyte systems, as well as the realization of scalable processing routes, to achieve commercially viable energy storage.

## REFERENCES

[1] M. M. Hasan et al., "Advancing energy storage: The future trajectory of lithium-ion battery technologies," *J. Energy Storage*, vol. 120, no. January, p. 116511, 2025, doi: 10.1016/j.est.2025.116511

[2] N. Lingappan, W. Lee, S. Passerini, and M. Pecht, "A comprehensive review of membrane membranes in lithium-ion batteries," *Renew. Sustain. Energy Rev.*, vol. 187, no. August, p. 113726, 2023, doi: 10.1016/j.rser.2023.113726

[3] S. Antony Jose et al., "Solid-state lithium batteries: Advances, challenges, and future perspectives," *Batteries*, vol. 11, no. 3, 2025, doi: 10.3390/batteries11030090

[4] C. Ikechukwu, U. Orji, A. P. I. Popoola, O. M. Popoola, and S. Mohammed, "A review of recent advances , current limitations , and remedies of lithium-ion batteries for advanced technological applications," *Futur. Batter.*, vol. 7, no. June, p. 100109, 2025, doi: 10.1016/j.fub.2025.100109

[5] S. Mahmud et al., "Recent advances in lithium-ion battery materials for improved electrochemical performance: A review," *Results Eng.*, vol. 15, no. May, p. 100472, 2022, doi: 10.1016/j.rineng.2022.100472

[6] Y. S. Chung, S. H. Yoo, and C. K. Kim, "Enhancement of meltdown temperature of the polyethylene lithium-ion battery Membrane via surface coating with polymers having high thermal resistance," *Ind. Eng. Chem. Res.*, vol. 48, no. 9, pp. 4346–4351, 2009, doi: 10.1021/ie900096z

[7] H. Ding et al., "Thermally stable poly-aromatic solid electrolyte coated polyethylene membrane as high-performance lithium-ion battery membrane," *J. Power Sources*, vol. 602, no. March, p. 234355, 2024, doi: 10.1016/j.jpowsour.2024.234355

[8] V. B. Mbayachi, "Ionic transport mechanisms in inorganic solid electrolytes: Interface, NMR and DNP studies," *Next Mater.*, vol. 8, no. April, p. 100657, 2025, doi: 10.1016/j.nxmate.2025.100657

[9] S. Janakiraman, A. Agrawal, R. Biswal, and A. Venimadhav, "An amorphous polyvinylidene fluoride-co-hexafluoropropylene based gel polymer electrolyte for sodium-ion cells," *Appl. Surf. Sci. Adv.*, vol. 6, p. 100139, 2021, doi: 10.1016/j.apsadv.2021.100139

[10] J. Moon, J. Y. Jeong, J. I. Kim, S. Kim, and J. H. Park, "An ultrathin inorganic–organic hybrid layer on commercial polymer separators for advanced lithium-ion batteries," *J. Power Sources*, vol. 416, pp. 89–94, 2019, doi: 10.1016/j.jpowsour.2019.01.058

[11] M. Shah and N. U. D. Hakim, "Advances in nanocellulose proton conductivity and applications in polymer electrolyte membrane fuel cells," *Next Mater.*, vol. 6, no. November 2024, p. 100484, 2025, doi: 10.1016/j.nxmate.2025.100484

[12] Z. Zou, Z. Hu, and H. Pu, "Lithium-ion battery membranes based-on nanolayer co-extrusion prepared polypropylene nanobelts reinforced cellulose," *J. Memb. Sci.*, vol. 666, no. September 2022, p. 121120, 2023, doi: 10.1016/j.memsci.2022.121120

[13] A. Gaddala, A. V. V. Sudhakar, S. Rafikiran, R. R. Deshmukh, and C. H. Basha, "Switched capacitor high voltage gain DC-DC Converters for hydrogen vehicles: A detailed comprehensive analysis," *e-Prime - Adv. Electr. Eng. Electron. Energy*, vol. 13, no. June, p. 101039, 2025, doi: 10.1016/j.prime.2025.101039

[14] Q. Liu, Z. Yu, Y. Hu, M. Tang, Q. Zhang, and M. Ye, "Designing advanced membranes toward lithium-ion batteries," *Adv. Energy Mater.*, vol. 03217, pp. 1–27, 2025, doi: 10.1002/aenm.202503217

[15] Y. Lei, J. Pei, P. Xue, L. Ma, Z. Liu, and R. Long, "Impacts of membrane on lithium dendrite growth and morphology under non-isothermal configurations," *J. Power Sources*, vol. 629, no. November 2024, 2025, doi: 10.1016/j.jpowsour.2024.236034

[16] A. Hahn, M. Ruf, S. Doose, and A. Kwade, "Enhancing lithium-ion battery safety: A comparative study of membrane performance under mechanical abuse," *Futur. Batter.*, vol. 6, no. March, p. 100064, 2025, doi: 10.1016/j.fub.2025.100064

[17] Y. Liu, S. Yu, Q. Shi, X. Ge, and W. Wang, "multilayer coatings for tribology: A mini review," *Nanomaterials*, vol. 12, no. 9, 2022, doi: 10.3390/nano12091388

[18] S. Kadanyo et al., "Enhancing compatibility and hydrophilicity of polysulfone/poly (ethylene-co-vinyl alcohol) copolymer blend ultrafiltration membranes using polyethylene glycol as hydrophilic additive and compatibilizer," *Sep. Purif. Technol.*, vol. 287, no. December 2021, p. 120523, 2022, doi: 10.1016/j.seppur.2022.120523

[19] Z. Fan et al., *Functionalized Membranes Boosting Electrochemical Performances for Lithium Batteries*, vol. 17, no. 1. Springer Nature Singapore, 2025. doi: 10.1007/s40820-024-01596-x

[20] S. Abrishami, H. Xiao, M. Asadnia, Z. X. Low, and A. Razmjou, "Recent advances in the design principles of lithium selective membranes," *Water Res.*, vol. 283, December 2024, 2025, doi: 10.1016/j.watres.2025.123724

[21] J. Lv, Q. Wang, M. OuYang, and Y. Cao, "Highly performing sodium metal batteries reinforced by a self-regulated dual-layered solid electrolyte interphase via a metal–organic framework," *ACS Appl. Mater. Interfaces*, vol. 16, no. 31, pp. 41570–41582, 23 Jul. 2024, doi: 10.1016/j.saa.2011.12.018

[22] G. Zhang, S. Wu, K. Li, B. Yuan, and Y. Zhang, "Advanced self-responsive thermal insulation coating: Protective strategies for thermal runaway fire in lithium-ion battery," *Chem. Eng. J.*, vol. 166898, 6 Aug. 2025, doi: 10.1016/j.ensm.2025.104233

[23] L. Spitthoff, P. R. Shearing, and O. S. Burheim, "Temperature, ageing and thermal management of lithium-ion batteries," *Energies*, vol. 14, no. 5, p. 1248, 25 Feb. 2021.

# 8 Membranes for Redox Flow Batteries (RFBs)

## 8.1 INTRODUCTION

Vanadium redox flow batteries (RFBs) are at the core of grid-scale energy storage, which we see as a solution to the variable output of renewables like solar and wind [1]. Unlike traditional batteries, which store energy within the cell, VRFBs use large external tanks of liquid electrolyte that contain vanadium ions for energy storage, and, during the power delivery stage, the electrolyte is passed through a flow cell [2]. It is the ion-selective membrane which is the key enabler in this design by allowing charge-balancing ions ($H^+$ or $SO_4^{2-}$) to pass through, but at the same time it blocks vanadium, which in its passage causes self-discharge, capacity fade, and efficiency loss. High ionic conductivity >100 mS/cm, which in turn minimizes ohmic losses, low permeability $<10^{-7}$ cm²/s for V-ions, and great performance under acidic (pH<1), oxidative ($V^{5+}$, and hygrothermal stresses >1000 cycles. Perfluorosulfonic acid (PFSA) membranes, which include Nafion, are the commercial dominant players due to robust proton conduction via hydrated nanochannels, but they also have a high price tag of \$500/m², vanadium permeability ($10^{-6}$ cm²/s), and issues with swelling-induced degradation. We look at Nafion composites, hydrocarbon cation-exchange membranes (SPEEK/PBI), anion-exchange membranes (AEMs), and hybrid aromatic ionomers through the lens of processing structure property relationships. Key performance indicators include vanadium permeability, area-specific resistance (ASR<0.2 Ω·cm²), and stack level efficiency, which should see >80% energy efficiency at 100 mA/cm². Also, we are seeing recent inroads with fillers modified hybrids (e.g., $SiO_2$/GO at 3–10 wt% and non-fluorinated designs which are reporting 2–5× selectivity improvements and 50% cost reduction), which are opening up paths to commercial scale [3].

## 8.2 MEMBRANE TYPES AND MATERIALS

The material for the membrane is the key factor that determines the effectiveness, durability, and also the economic part of the VRFBs. A variety of ion-conducting membranes have been covered, and each one of them has different pros and cons.

### 8.2.1 Perfluorosulfonic Acid Membranes

PFSA membranes are widely used in fuel cells and electrochemical devices because they offer excellent proton conductivity, chemical stability, and mechanical strength. The most common PEM material is poly(tetrafluoroethylene-co-perfluorosulfonic acid), like Nafion. The hydrophobic PTFE main chain contributes to the membrane's


DOI: 10.1201/9781003679363-8

mechanical strength. The hydrophilic perfluorinated side chains bearing sulfonic acid groups ($SO_3H$) cause phase separation [4]. This forms ionic channels that enable high proton conductivity in the hydrated membrane by forming hydronium ions ($H_3O^+$) via Grotthuss hopping. More ionic groups boost conductivity but also increase water absorption, leading to swelling that harms mechanical stability. To counter this, researchers have incorporated inorganic fillers like $SiO_2$, $TiO_2$, $ZrO_2$, metal phosphates, heteropolyacids (HPAs), zeolites, clays, and functionalized graphene oxide. These improve water retention, scavenge free radicals, and enhance high-temperature performance [5]. Recent trends focus on hybrid composites in which fillers interact with $SO_3H$ sites via dynamic cross-links. This reduces chain mobility, increases modulus, and allows operation up to 120–135°C with lower methanol permeability and better selectivity. For instance, Nafion-$SiO_2$ hybrids possess enhanced thermal stability, anisotropic diffusion properties, and conductivity at low relative humidity. Nafion-ZrP or $TiO_2$ derivatives possess superior water uptake properties with reduced gas crossover properties. Reinforced multilayer materials include ePTFE-impregnated Nafion XL with PTFE mesh stabilizers, which reduce in-plane swelling by an order of magnitude. They boost yield strength, fracture energy, and lifespan under humidity cycling, often outperforming recast Nafion in PEFC power density and OCV retention [6].

### 8.2.2 Modified Nafion Composite Membranes

Modified Nafion composite membranes are developed by incorporating inorganic nanomaterials into the Nafion matrix to overcome its inherent limitations, such as high methanol crossover, poor water retention at elevated temperatures, excessive swelling, and chemical instability. These modifications significantly enhance proton conductivity, mechanical strength, thermal stability, and the overall performance of direct methanol fuel cells (DMFCs). Nafion consists of a hydrophobic PTFE backbone that provides mechanical strength, while the hydrophilic sulfonic acid ($SO_3H$) groups form ionic channels that allow proton transport. Proton conduction mainly occurs through the Grotthuss mechanism involving hydrogen bond breaking and formation, and through the vehicular diffusion of hydronium ions under hydrated conditions. However, Nafion exhibits high methanol permeability greater than $10^{-7}$ $cm^2/s$ at 60°C, and its conductivity decreases above 80°C, which restricts its use in DMFC systems operating with high methanol concentrations such as 9 M [7].

To address these limitations, various modification strategies have been explored using different types of inorganic fillers. These include solid nanoparticles such as $SiO_2$, $TiO_2$, $ZrO_2$, and $Al_2O_3$ hollow or porous materials such as SBA-15 silica and metal organic frameworks like UiO-66 and ZIF-8, one-dimensional structures such as halloysite nanotubes and $TiO_2$ nanofibers, two-dimensional materials such as graphene oxide, layered double hydroxides, hexagonal boron nitride, and montmorillonite clay, three-dimensional porous frameworks including MOFs and zeolites, and natural clay materials like sepiolite and montmorillonite. These fillers are typically introduced through fabrication methods such as solution casting using doctor blade or Petri dish techniques, sol–gel synthesis, dip coating, electrospinning to produce non-woven nanofiber mats, chemical grafting, and layer-by-layer assembly [8].

The incorporation of these materials results in the formation of tortuous diffusion paths, hydrogen-bonded networks, and controlled pore structures, which are effective in the suppression of methanol crossover while maintaining efficient proton transport. For example, $SiO_2$ fillers with loadings between 3 and 10 wt% prepared through sol–gel techniques can increase water uptake up to 25% while enhancing proton conductivity to around 39–242 mS/cm at 80°C. At the same time, methanol permeability can decrease to approximately 5–27 × $10^{-7}$ $cm^2/s$, which is 20–70% lower than that of recast Nafion. These improvements can result in power densities that are 1.4 to 2 times higher, with values around 117–128 $mW/cm^2$ at operating temperatures between 60 and 80°C. Similarly, $TiO_2$ nanoparticles with concentrations between 0.5 and 10 wt% improve membrane selectivity up to 187 × $10^4$ $Scm^{-3}$ and enhance power density in the range of 44–135 $mW/cm^2$ due to strong acid-base interactions with the sulfonic acid groups. Membranes modified with zirconium phosphate, using a modification weight percentage range of 8–26% and a dip coating method, are capable of reducing methanol crossover to a rate of 7.25 × $10^{-7}$ $cm^2/s$ while maintaining a conductivity of 13 mS. Graphene oxide and sulfonated graphene oxide fillers at concentrations between 0.5 and 1.8 wt% form bipolar barrier structures that significantly suppress methanol diffusion and enable conductivities between 214 and 262 mS/cm with power densities reaching nearly 300 $mW/cm^2$ and permeability reductions of 50–85% [9].

Additionally, clay-based fillers such as montmorillonite and sepiolite at a level of 10 wt% contribute to the improvement of the membrane's performance through the development of tortuous paths for transport and hydrogen bonding interactions. This is due to the improvement of the ion-exchange capacity at a level of 0.1–2.01 meq/g, as well as the proton conductivity at a level of 14–240 mS/cm. Porous materials such as MOFs and zeolites, including mordenite at around 10 wt%, utilize their internal pore networks and acidic sites to achieve conductivities of 12–46 mS/cm while reducing methanol crossover by nearly 85%. In addition, electrospun non-woven nanofiber structures such as $TiO_2$–SPEEK nanofibers provide highly porous membranes with proton conductivities between 100 and 240 mS/cm and power densities in the range of 42–86 $mW/cm^2$, along with tensile strengths approaching 942 MPa. In most cases, the optimal filler concentrations ranging from 3 to 10 wt% are considered to have the best compromise between conductivity durability and mechanical stability. At these loadings, composite membranes demonstrate improved durability, such as stable operation for up to 50 hours at temperatures around 130°C and enhanced oxidative stability with more than 2.5 wt% membrane retention in Fenton's reagent tests. Too much filler material could result in particle agglomeration, which might affect the performance, but the addition of sulfonic and amino groups could facilitate the dispersion and interaction with the Nafion matrix. As a result, these modified composite membranes show significantly improved performance compared with pure Nafion, especially in DMFC systems operating at high methanol concentration and elevated temperatures [10].

### 8.2.3 Hydrocarbon-Based Cation-Exchange Membranes

Hydrocarbon-based cation-exchange membranes, mainly composed of aromatic sulfonated polymers such as sulfonated poly(arylene ether ketone) (SPEEK), sulfonated poly(ether ether ketone) (sPEEK), and polybenzimidazole (PBI) blends, are emerging as cost-effective fluoro-free alternatives to PFSA (Nafion) proton exchange membranes used in fuel cells. These materials provide tunable ion-exchange capacity in the range of 0.8–2.5 meq/g along with high thermal stability above 450°C and low methanol permeability, which helps address the high cost and environmental concerns associated with PFSA membranes [11]. The polymer's structure consists of a rigid aromatic backbone with ether ketone linkages and pendant sulfonic acid groups (–$SO_3H$), leading to microphase separation and ionic domains for proton transport via Grotthuss and vehicular mechanisms in a hydrated state. In polymer blends such as SPEEK–PBI, proton transport is further enhanced through acid-base interactions that form ionic cross-links, which reduce free volume and swelling. Pristine sPEEK membranes show relatively high proton conductivity in the range of 0.07–0.21 S/cm at temperatures between 80 and 100°C, but suffer from high water uptake above 65 wt% at 80°C which causes dimensional instability. Combining sPEEK with PBI at 5–20 wt% by solution casting in DMSO limits swelling to 9–21 wt% at 80°C. The mechanical properties show increased values of 22 to 37 MPa in tensile strength and 0.51 to 0.82 GPa in modulus. The blends also show improved oxidative stability of more than 6–10 hours in Fenton's reagent tests and lower methanol permeability in the range of 1.4–5.3 × $10^{-7}$ $cm^2/s$, which is nearly six times lower than that of Nafion, thus providing selectivity values around two times higher than Nafion while maintaining proton conductivity between 0.026 and 0.145 S/cm at 100°C [12]. The synthesis process is usually carried out via mild sulfonation of phenylated PEEK at room temperature using $H_2SO_4$, where pendant -$SO_3H$ groups are introduced, while PBI is synthesized via the direct polycondensation of diacids and tetraamines in a polyphosphoric acid medium, while the membranes are prepared via blending, where the two components are miscible due to the hydrogen bonding interaction between the imidazole groups and the sulfonic acid groups, as proved by the shift of the FTIR bands from 1,290 to 1,281 $cm^{-1}$, along with the results obtained via NMR Thermogravimetric analysis shows a two-step degradation pattern where the first stage corresponds to $SO_3H$ group loss around 250°C and the second stage represents backbone degradation above 450°C with Td10 values between 388 and 435°C for blends containing about 20 wt% PBI which indicates better thermal stability than Nafion [13]. These membranes perform effectively in DMFC and PEMFC systems operating at elevated temperatures between 80 and 100°C, with blend-10 providing the best balance between conductivity, methanol barrier properties, and mechanical stability, and showing selectivity and performance comparable to Nafion, which makes these hydrocarbon-based membranes promising candidates for durable and high-performance fuel cell applications [14].

### 8.2.4 Anion-Exchange Membranes

AEMs are solid polymer electrolytes used in alkaline fuel cells where $OH^-$ ions move from the cathode to the anode [15]. Compared with proton exchange membrane fuel cells, they offer advantages such as faster non-PGM catalyst kinetics using materials like Ag, Co, or Ni, lower platinum usage, greater fuel flexibility, including hydrogen and alcohols, and reduced fuel crossover due to opposite electro-osmotic drag. However, they still face limitations such as lower ionic conductivity, typically around 20–100 mS/cm, compared with Nafion, which exceeds 100 mS/cm, along with issues related to alkaline degradation and $CO_2$ carbonation that forms $HCO_3^-$ and $CO_3^{2-}$ species, which can reduce conductivity by nearly 70–80% [16]. Commercial AEMs such as Tokuyama A201 with conductivity around 42 mS/cm, Fumasep FAA-3 with about 40–45 mS/cm, and Sustainion with about 70 mS/cm generally use polystyrene-DVB or hydrocarbon backbones containing cationic functional groups such as quaternary ammonium, imidazolium, phosphonium, or spirocyclic cations to enable $OH^-$ ion exchange [17]. These membranes are typically synthesized through methods such as radiation or chemical grafting, for example, ETFE or LDPE grafted with vinylbenzyl chloride followed by trimethylamine functionalization, quaternization reactions, or ring opening metathesis polymerization, copolymerization, while targeting ion-exchange capacities around 1.5–2.5 meq/g, membrane thickness between 20 and 50 μm and chemical stability in 1–10 M KOH at temperatures between 60 and 80°C [18]. Polyvinyl alcohol-based AEMs cross-linked with glutaraldehyde and doped with KOH or blended with polymers such as PDDA, PEI, or PAA can achieve conductivity values between 0.03 and 0.31 S/cm but often exhibit high swelling (>50 wt%) and limited durability [19]. Membranes based on polystyrene or polysulfone, prepared by chloromethylation followed by functionalization with DABCO or quaternary ammonium groups, offer improved rigidity, alkaline resistance, and ion-exchange capacity of 1–2 meq/g, conductivity of 0.02–0.1 S/cm, and tensile strength >30 MPa. Polyphenylene or PPO-based membranes functionalized with quaternary ammonium or imidazolium groups, often with spacer chains, show higher conductivity between 50 and 110 mS/cm at 80°C with lower swelling around 10–30% due to well-defined phase-separated structures [20]. Polyolefin membranes produced through radiation grafting of LDPE, ETFE, or FEP with vinylbenzyl chloride, followed by trimethylamine functionalization, can reach conductivity values between 50 and 320 mS/cm, while polymeric ionic liquids-based membranes containing imidazolium ionic liquid structures typically show conductivity between 10 and 73 mS/cm and good $CO_2$ tolerance in thin film forms. Hybrid composite membranes, where nanofillers like $SiO_2$, $ZrO_2$, or graphene oxide are dispersed in the matrix, such as PVA, PSU, or PPO, are known to improve the conductivity as well as the stability of the membranes, with conductivity ranging from 50 to 215 mS/cm, along with a retention of nearly 90% after 500 hours of exposure to alkaline conditions. Recent AEMFC systems have demonstrated peak power densities between 1 and 3.4 W/cm$^2$ using hydrogen and oxygen with low platinum or PGM-free catalysts while future development targets from the DOE aim for 1 W/cm$^2$ using PGM-free catalysts and around 600 mW/cm$^2$ using air which requires membranes with conductivity above 100 mS/cm $CO_2$-related voltage losses below 40 mV and durability exceeding 1000 hours at 80°C and 70% relative humidity.

### 8.2.5 Hybrid Aromatic Ion-Exchange Membranes

AIEMs are obtained as conductivity of 0.1–0.46 S/cm at 80°C for many systems, but oxidative degradation remains a concern. Modifications such as GO-g-SPEEK or Nafion blends with conductivity around 0.23 S/cm, EG or CNC crosslinking with conductivity about 0.186 S/cm, PVDF nanofiber reinforcement giving DMFC power density around 104 mW/cm², and core–shell SSi-GO nanofibers with methanol permeability about 39% of Nafion improve retention and selectivity. These membranes typically show IEC values around 1.25–2.16 meq/g with conductivity between 0.13 and 0.186 S/cm, but hydrolytic weakness can appear. This limitation is reduced through approaches such as aliphatic or perylenediimide chains, giving power density up to 931 mW/cm², side chain $-SO_3H$ functionalization giving DMFC power density about 28.7 mW/cm², and nanofiber frameworks where conductivity exceeds Nafion NR212 and RHA biofillers with conductivity around 0.206 S/cm. In PSF or SPES systems with DS values between 19 and 100 conductivity ranges from about 0.01 to 0.35 S/cm and can be improved using chitin nanowhiskers with water uptake around 19% S-MCM-41 fillers giving conductivity near 0.069 S/cm serine-CW composites reaching conductivity around 0.234 S/cm with power density about 74 mW/cm² Cr-MIL-101-$NH_2$ MOF systems with conductivity about 0.041 S/cm at high temperature and power density around 238 mW/cm² and g-$C_3N_4$ fillers with conductivity around 0.12 S/cm at 180°C. For PSIs phosphoric acid-doped membranes with ADL values of 5–27 show anhydrous conductivity between 0.02 and 0.235 S/cm at temperatures around 160–200°C with power density between 250 and 690 mW/cm² and further improvements are observed with benzimidazole grafting giving conductivity about 0.212 S/cm polymeric ionic liquid crosslinking giving about 0.117 S/cm BA-a branching achieving about 690 mW/cm² with stable OCV for 200 h and PGO based systems reaching about 359 mW/cm². PPO-based membranes, such as SPPO or QPPO, show conductivity between 0.02 and 0.085 S/cm with good methanol resistance, producing about 81 mW/cm² in DMFC with HGM fillers, while imidazolium functionalization gives about 280 mW/cm² at high temperature, and triazole crosslinked systems show conductivity around 0.064 S/cm at 180°C with hybrid phosphonic siloxane structures reaching about 638 mW/cm². Overall strategies including crosslinking such as EG or BA-a nanofillers such as GO MOF $SiO_2$ or CNC polymer blending such as SPEEK-BPPO block copolymers such as PAEK-b-KSPAEK and nanofiber frameworks help balance conductivity which can reach up to about 1.4 times that of Nafion while maintaining water uptake between 20 and 136% mechanical strength around 0.4–1.3 GPa and stability of about 10 h in Fenton's test and 200–2,000 h in MEA operation leading to power densities between about 100 and 931 mW/cm² which in several cases outperform Nafion particularly under high methanol concentration or low humidity conditions [1].

# 8.3 ROLES AND KEY CHARACTERISTICS OF MEMBRANE IN RFBS

## 8.3.1 Membrane–Electrolyte Interactions

Membrane–electrolyte interaction refers to the chemical and physical interactions between water ions and the polymer matrix in ion-exchange membranes, including PFSA membranes such as Nafion. Water molecules are first attracted to the sulfonic acid groups ($–SO_3H$) through a hydrogen bonding mechanism, resulting in the formation of hydronium ions like $H_3O^+$ and even larger clusters like $H_9O_4^+$ in the primary hydration shells consisting of 1–6 water molecules per ionic site. Significant amounts of solvation energy are released in this initial stage of hydration, in the range of 20–25 kJ/mol. The bound water then forms hydrophilic nanodomains within the hydrophobic PTFE backbone, which resists excessive swelling through elastic deformation and establishes a chemical and mechanical equilibrium that determines the overall water uptake and phase separation in the membrane. For a level of hydration above a certain threshold, water molecules per $–SO_3^-$ group, the additional water molecules are seen to behave in a similar way to bulk water and thus create larger pools in which continuous percolation pathways are possible. At the same time, surface reorganization occurs where the membrane shifts from a hydrophobic state under dry or vapour conditions to a more hydrophilic structure in the presence of liquid water. In AEMs, quaternary ammonium or imidazolium cations coordinate with $OH^-$ ions to enable ion transport, but $CO_2$ contamination can rapidly react with $OH^-$ to form bicarbonate ($HCO_3^-$) and carbonate ($CO_3^{2-}$) species, which occupy the ionic channels and reduce ion mobility, leading to a conductivity decrease of about 70–80% due to altered hydration structure and restricted ion movement [21]. The influence of electrolyte composition on the key physicochemical and electrochemical properties of membranes is summarized in Table 8.1.

**TABLE 8.1**
**Effect of Electrolyte Composition on Membrane Performance**

| Electrolyte Composition | Conductivity Trend | Swelling Behaviour | Crossover Tendency | Results |
|---|---|---|---|---|
| 2 M $H_2SO_4$ | High conductivity | Moderate swelling | Moderate crossover | Standard electrolyte for VRFBs. |
| 4 M $H_2SO_4$ | Very high conductivity | Excessive swelling | High crossover | Accelerates degradation. |
| 2 M $H_2SO_4$ + 1 M $H_3PO_4$ | Moderate conductivity | Low swelling | Reduced crossover | Improved stability; used in hybrid electrolytes. |
| 2 M $H_2SO_4$ + 0.5 M HCl | High conductivity | Slightly high swelling | Moderate crossover | Chloride ions may attack polymer chains. |

### 8.3.2 Ion Transport Mechanisms in RFB Membranes

In AEM-RFB systems, $OH^-$ transport occurs mainly through hydrogen-bonded networks in bound water, which dominates at low hydration levels, as shown in Figure 8.1. When hydration increases, hydrated ions begin to diffuse through hydrophilic channels about 3–6 nm in size, which form continuous percolation pathways once a threshold is reached. Electro-osmotic drag carries about 1–3 water molecules per ion towards the cathode, and in AEMs, water typically moves from anode to cathode, which helps reduce vanadium crossover. This process is balanced by back diffusion driven by water activity gradients. Under confined conditions, the narrow ion channels restrict bulk diffusion but improve ion selectivity, such as $H^+$ over $VO^{2+}$. Additional coupled transport phenomena can also occur, including thermo-osmosis, which is driven by temperature gradients and streaming potentials generated by pressure or concentration gradients. Incorporation of fillers such as $SiO_2$ or MOFs increases the tortuosity of transport pathways, which can enhance conductivity to about 1.4 times that of Nafion while reducing species crossover by nearly 70%. In PFSA membranes, vehicular transport dominates at high relative humidity, while the Grotthuss mechanism becomes more important at lower humidity levels. In AEM-RFB systems, the direction of electro-osmotic drag is reversed, which helps minimize anode flooding, though it may increase the risk of cathode dehydration. Overall membrane performance is typically observed in the conductivity range of 50–150 mS/cm, where the balance between conductivity and selectivity is controlled by equivalent weight and membrane morphology [22].

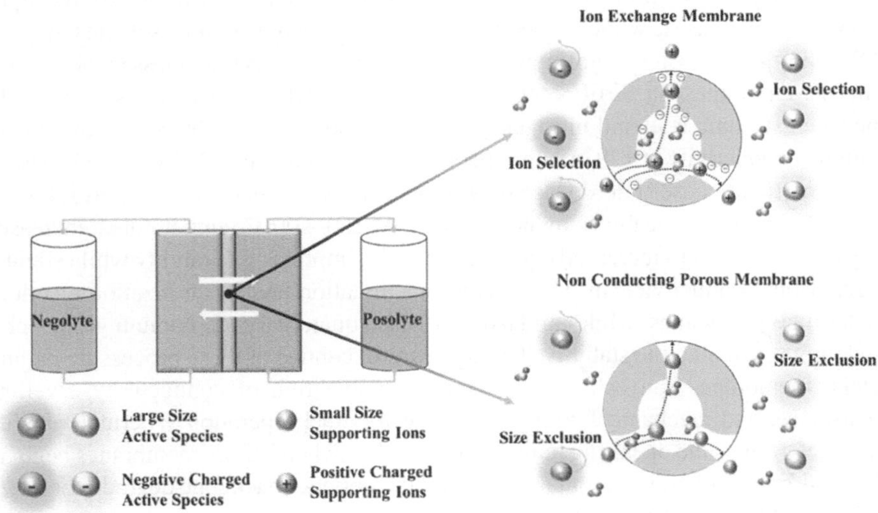


**FIGURE 8.1** Schematic demonstration of an ion-selective membrane for a redox flow battery.

### 8.3.3 Interfacial Phenomena and Contact Resistance

PFSA membrane surfaces change significantly upon hydration, where the contact angle decreases from about 90° to nearly 30°. This transformation is the cause of vapour–liquid hysteresis. It may also result in increased area-specific resistance (ASR > 0.1 $\Omega \cdot cm^2$) due to poor wetting of the catalyst layer. The use of ionomer thin films, such as 20 wt% Sustainion, helps fill interfacial gaps and can reduce impedance by nearly 50%. Stack pressurization also affects the membrane–electrode interface, where compression alters the spacing between ionic domains as observed from SAXS peak shifts, which results in a nonlinear increase in contact resistance. Repeated swelling and shrinking during operation can cause interfacial delamination and loss of structural integrity, but reinforcement strategies such as nanofiber structures like SPEEK–PVDF mats or pore-filling membrane designs improve adhesion and mechanical stability. Electrochemical impedance spectroscopy normally displays a semicircle corresponding to charge transfer resistance and contact resistance. Optimum performance occurs at a compression pressure between 0.5 and 1 MPa. Overall, interfacial effects contribute roughly 20–40% of the total ASR and can be controlled through factors such as solvent selection during membrane casting and the use of asymmetric ionomer structures [23].

### 8.3.4 Process–Structure–Property Relationships

Process–structure–property relationships link fabrication parameters to membrane performance. Increasing the sulfonation degree (DS 0.8–2.5) raises the ion-exchange capacity (IEC 0.9–2.0 meq/g) and conductivity but can also cause excessive swelling. In extrusion casting of PFSA membranes, crystalline domains develop, and the alpha relaxation increases, while dispersion casting using high boiling solvents such as DMF produces greater chain entanglement, which improves toughness by nearly two times. Membranes like SPEEK 60 wt% with 40 wt% PBI exhibit acid-base cross-links between sulfonic acid and nitrogen groups, reducing swelling by 30% while maintaining conductivity at 120 mS/cm. Incorporation of fillers like 1–5 wt% GO CNC or MOF particles causes microphase separation, enhancing conductivity up to 1.4 times while reducing permeability by nearly 70%. At 150–200°C temperatures, increased crystallinity leads to increased equivalent weight, improving durability while slightly decreasing conductivity. In AEMs, chloromethylation and quaternization reactions determine IEC values, while imidazolium-based quaternary ammonium groups provide improved alkaline stability. Through careful control of these processing parameters, membranes can achieve a balanced combination of conductivity, swelling resistance, and mechanical strength suitable for stack operation. Thermal stability also determines the operating temperature window where PFSA membranes begin to lose sulfonic acid groups around 250°C and the polymer backbone degrades between 450 and 500°C. Under vapour equilibrium conditions, hydration levels decrease with increasing temperature due to entropic clustering and surface effects, whereas in liquid environments, this results in a lower modulus and increased swelling. In practice, the optimal operating range for PFSA membranes is about 80–120°C before significant dehydration occurs, which can reach nearly 90%. Membranes reinforced with PFSA resins like Gore Select, which contain PTFE mesh, show good stability at

**TABLE 8.2**
**Influence of Membrane Parameters on Stack Performance**

| Parameter | Impact on Stack Performance |
|---|---|
| **Thickness** | Thinner membranes reduce ohmic losses but risk pinholes and mechanical failure. |
| **Compressibility** | Affects gasket sealing; too soft leads to leakage, too rigid leads to poor conformity. |
| **Mechanical Strength** | Prevents rupture under pressure gradients, ensuring long-term stability. |
| **Dimensional Stability** | Controls swelling/shrinkage, which impacts sealing integrity and resistance. |
| **Surface Compatibility** | Enhances adhesion with electrodes and reduces interfacial resistance. |
| **Chemical Durability** | Enables prolonged exposure to acidic and oxidative environments. |
| **Cost & Scalability** | Determines commercial viability; roll-to-roll production reduces cost. |

130°C and 50% relative humidity for 500 hours. Hydrocarbon-based membranes like SPEEK have a glass transition temperature of 220°C. Phosphoric acid-doped PBI is found to have good performance under anhydrous conditions with conductivity as high as 0.2 S/cm at 180°C. Increasing temperature initially enhances conductivity but eventually leads to a plateau and then a decline due to dehydration and structural changes in the membrane. Table 8.2 outlines the influence of critical membrane parameters on stack performance and operational efficiency.

### 8.3.5 Thermal Stability and Operating Temperature Effects

PFSA membranes show thermal decomposition through $SO_3$ group release in the range of about 250–400°C, followed by polymer backbone breakdown near 450°C. The $Td_{10}$ value generally increases with higher crystallinity obtained through annealing, although radical formation can accelerate degradation. Above 60°C, an increase in conductivity is often seen as the hydrophilic channels expand, but at the same time, crossover can increase as Donnan exclusion and osmotic effects decrease. In some instances, the best performance is seen at 40–50°C, as conductivity is maintained while crossover and chemical degradation are minimized. Electrolyte viscosity also decreases with increasing temperature, which can create pressure gradients that enhance convective transport. For systems such as vanadium RFBs, operation is typically kept below 60°C to prevent rapid capacity loss and maintain long-term stability [24].

### 8.3.6 Membrane Integration in Stack Design

In the stack integrations, very thin membranes with a thickness below 20 μm significantly reduce ohmic loss, with IR drop reduced by nearly 60%. PTFE mesh reinforcements also help to avoid pinholes and crossover for large cells above 500 cm². Membrane–electrode assemblies usually employ ionomers with a weight % between 20 and 50 wt%, with low molecular weight for improved penetration into the triple phase boundary. Asymmetric designs, for instance, high IEC cathode AEI layers, allow the use of PGM-free cathodes with power densities above 3 W/cm², achievable for $H_2/O_2$ systems. Hybrid membrane materials such as Nafion/ZrP with 10 wt% filler content retain nearly 80% of their conductivity after 1000 cycles at 130°C. Low-permeability flow field designs also help to minimize shunt currents for stacked cells. Nanofiber or block copolymer membrane structures also enhance methanol permselectivity by nearly five times, with power densities of 931 mW/cm² achievable. In AEM stacks, water management is important because water is produced at the anode and consumed at the cathode, and some systems show self-purging behaviour for $CO_2$. With reinforced ionomer membranes, the stack durability of up to about 2000 hours at current densities around 0.5 A/cm² has been demonstrated.

## 8.4 PERFORMANCE LIMITATIONS IN VRFBS

Vanadium RFBs often experience performance limitations due to high ionic resistance in the membrane and vanadium ion crossover. Vanadium permeability typically lies in the range of about $10^{-7}$–$10^{-6}$ cm²/s, which leads to self-discharge rates of roughly 1–2% per day. Ion crossover mainly occurs through transport mechanisms across the membrane, such as diffusion, convection, and electro-osmotic drag. Osmotic transport driven by pressure differences and electro-osmotic flow associated with ion movement both contribute to the transfer of vanadium species between half-cells. This effect becomes more pronounced when there is a viscosity difference between the positive and negative electrolytes at constant flow rates, which results in a net vanadium migration across the membrane. Over repeated cycling, this imbalance can cause significant capacity fading, with studies reporting nearly 13% capacity loss after around 35 cycles. Sulfonic group ($SO_3$) loss occurs in a range of 250–400°C, and after this, backbone oxidation occurs above 450°C. Hydrocarbon membranes are more selective, and in this case, conductivity also reduces. Side chain unzipping occurs during degradation, not main chain scission. The fluoride release rate (FRR) is used as an indicator of membrane lifetime and is influenced by $H_2O_2$ crossover during operation, which accelerates degradation.

## 8.5 FUTURE PERSPECTIVES ON RFB MEMBRANES

### 8.5.1 VRFB Membrane and Operation

Validate asymmetric flow experimentally, where models report that continuous pressure conditions, meaning higher flow on the more viscous side, can lead to a capacity loss of about 90%. Full-scale tests with real electrolytes are also reported. Viscosity optimized electrolytes should be developed with a parallel design of viscosity to reduce or eliminate osmotic convection, and can be paired with thin low convection membranes such as thin PFSA [25].

### 8.5.2 PFSA Ionomer Improvements

Side chain or EW hybrid designs for thin films study behaviour at the 100 nm scale in electrodes where substrate interactions change morphology and transport, and develop nano to macro scale models for prediction. Multiscale electro-chemical-mechanical models integrate the effects of chemical degradation, such as $OH^-$ induced unzipping, mechanical wear from hygrothermal cycling, and transport behaviour under in situ conditions. Redox flow selectivity focuses on quantifying multi-ionic transport, such as $V^{4+}$ and $H^+$ in HBr or VRFB electrolytes and developing Donnan-tuned PFSA membranes.

### 8.5.3 Nafion Composites for DMFC

Prevention of Agglomeration: This involves the functionalization of 2D or 3D fillers such as GO, MOFs, and LDHs with polymer materials to ensure the uniform distribution of the fillers at concentrations above 10 wt%. It is important to prevent the blocking of the transport channels. Optimization of Electrospinning: This is done by controlling humidity and temperature to produce uniformly sized nanofibers. Small, sharp-edged particles are tested to prevent the formation of rough surfaces. MOFs Stability: This is done by using MOFs that are stable under acidic conditions, such as UiO-66-NH2, which is stable under the SPEEK or Nafion environment. This is important for scaling up the commercial application of DMFCs. Coating Adhesion: This is done by using the growth-based method for coating nanoparticle thin films, as well as the hybrid sol–gel method for creating porous structures.

### 8.5.4 Hybrid PEMs

Durability under real stacks requires long-term MEA tests (1000 $h^+$) that include fluctuating temperature, pressure, and fuel conditions, while monitoring DS degradation in SPEEK/SPBI. Commercial scale-up focuses on optimizing the post-sulfonation process of PEEK/PSF to reduce cost and using nanofiber or block copolymers for controlled phase separation. HT-PEMFCs (>100°C) use PA-doped PBI with an imidazole–triazole balance and anhydrous conductors through ionic liquid functionalization. In-operando diagnostics employ neutron and X-ray imaging to monitor morphological changes under stress. AI/ML modelling is used to understand filler-polymer interactions within composites. Sustainability focuses on recycled materials in fuel cells and bio-based fillers for greener PEMs.

## 8.6 CONCLUSION

Membranes are the core functional component of VRFBs, and their role goes far beyond simple ion separation. They define ionic transport behaviour, ion selectivity, chemical stability, as well as long-term electrochemical durability under the strongly acidic and oxidative conditions characteristic of the operation of a VRFB. While perfluorosulfonic acid-based membranes like Nafion represent the standard for performance, their high cost as well as their relatively high vanadium ion permeability have triggered considerable effort to find substitutes that provide a better balance of ion selectivity, conductivity, and cost. Developments in membrane technology have shown considerable promise in the design of novel membranes that include the use of nanomaterials, the addition of barrier layers, as well as hybrid ion-exchange materials. These advancements have the potential to improve the performance-cost ratio of the VRFB system to allow for the use of higher current densities while minimizing crossover effects. Roll-to-roll processing is increasingly becoming critical in the development of membrane technologies as the field transitions from the laboratory to the industrial scale. Future progress in VRFB membrane development will rely on the combined integration of molecular design, structural engineering, and scalable processing techniques. Successfully bringing these aspects together is expected to produce membranes that are efficient, durable, and economically competitive, thereby supporting the wider deployment of VRFB technology for large-scale energy storage of intermittent renewable energy sources.

## REFERENCES

[1] A. Leba Akman, M. Z. Arslan, and M. Farsak, "The rise of vanadium redox flow batteries: A game-changer in energy storage," *J. Alloys Compd.*, vol. 1038, no. August, p. 182869, 2025, doi: 10.1016/j.jallcom.2025.182869

[2] T. T. K. Huynh, T. Yang, P. S. Nayanthara, Y. Yang, J. Ye, and H. Wang, *Construction of High-Performance Membranes for Vanadium Redox Flow Batteries: Challenges, Development, and Perspectives*, vol. 17, no. 1. Springer Nature Singapore, 2025.

[3] K. W. Knehr and E. C. Kombu, "Electrochemistry Communications Role of convection and related effects on species crossover and capacity loss in vanadium redox flow batteries," *Electrochem. Commun.*, vol. 23, pp. 76–79, 2012, doi: 10.1016/j.elecom.2012.07.008

[4] M. Cecchetti, T. A. Ebaugh, L. Bonville, R. Maric, A. Casalegno, and M. Zago, "Development of a vanadium redox flow battery operating with thin membrane coupled with a highly selective and stable silica-based barrier layer," *Energy Technol.*, vol. 12, no. 8, pp. 1–12, 2024, doi: 10.1002/ente.202400410

[5] S. Bukola et al., "Single-layer graphene as a highly selective barrier for vanadium crossover with high proton selectivity," *J. Energy Chem.*, vol. 59, pp. 419–430, 2021, doi: 10.1016/j.jechem.2020.11.025

[6] P. Feng et al., "Novel sulfonated poly(ether ether ketone)/polybenzimidazole blends for proton exchange membranes," *High Perform. Polym.*, vol. 25, no. 6, pp. 697–704, 2013, doi: 10.1177/0954008313482956

[7] J. Du, H. Lin, L. Zhang, and L. Wang, "Advanced materials for vanadium redox flow batteries: Major obstacles and optimization strategies," *Adv. Funct. Mater.*, vol. 2501689, pp. 1–35, 2025, doi: 10.1002/adfm.202501689

[8] H. Y. Jung, M. S. Cho, T. Sadhasivam, J. Y. Kim, S. H. Roh, and Y. Kwon, "High ionic selectivity of low permeable organic composite membrane with amphiphilic polymer for vanadium redox flow batteries," *Solid State Ionics*, vol. 324, no. February, pp. 69–76, 2018, doi: 10.1016/j.ssi.2018.06.009

[9] M. Shoaib et al., "Advances in redox flow batteries – A comprehensive review on inorganic and organic electrolytes and engineering perspectives," *Adv. Energy Mater.*, vol. 14, no. 32, 2024, doi: 10.1002/aenm.202400721

[10] P. Zuo et al., "Ion exchange membranes: Constructing and tuning ion transport channels," *Adv. Funct. Mater.*, vol. 32, no. 52, pp. 1–37, 2022, doi: 10.1002/adfm.202207366

[11] W. W. Ng, H. S. Thiam, Y. L. Pang, K. C. Chong, and S. O. Lai, "A state-of-art on the development of Nafion-based membrane for performance improvement in direct methanol fuel cells," *Membranes (Basel).*, vol. 12, no. 5, 2022, doi: 10.3390/membranes12050506

[12] S. N. Oreiro, A. Bentien, J. Sloth, M. Rahimi, M. B. Madsen, and T. Drechsler, "Crossover analysis in a commercial 6 kW/43kAh vanadium redox flow battery utilizing anion exchange membrane," *Chem. Eng. J.*, vol. 490, no. May, p. 151947, 2024, doi: 10.1016/j.cej.2024.151947

[13] I. Derr, M. Bruns, J. Langner, A. Fetyan, J. Melke, and C. Roth, "Degradation of all-vanadium redox flow batteries (VRFB) investigated by electrochemical impedance and X-ray photoelectron spectroscopy: Part 2 electrochemical degradation," *J. Power Sources*, vol. 325, pp. 351–359, 2016, doi: 10.1016/j.jpowsour.2016.06.040

[14] D. I. Kushner, A. R. Crothers, A. Kusoglu, and A. Z. Weber, "Transport phenomena in flow battery ion-conducting membranes," *Curr. Opin. Electrochem.*, vol. 21, pp. 132–139, 2020, doi: 10.1016/j.coelec.2020.01.010

[15] R. Hammer, M. Schönhoff, and M. R. Hansen, "Comprehensive picture of water dynamics in nafion membranes at different levels of hydration," *J. Phys. Chem. B*, vol. 123, no. 39, pp. 8313–8324, 2019, doi: 10.1021/acs.jpcb.9b05093

[16] S. Maurya, S. H. Shin, Y. Kim, and S. H. Moon, "A review on recent developments of anion exchange membranes for fuel cells and redox flow batteries," *RSC Adv.*, vol. 5, no. 47, pp. 37206–37230, 2015, doi: 10.1039/c5ra04741b

[17] R. Banerjee, N. Bevilacqua, L. Eifert, and R. Zeis, "Characterization of carbon felt electrodes for vanadium redox flow batteries – A pore network modeling approach," *J. Energy Storage*, vol. 21, no. September 2018, pp. 163–171, 2019, doi: 10.1016/j.est.2018.11.014

[18] R. S. L. Yee, K. Zhang, and B. P. Ladewig, "The effects of sulfonated poly(ether ketone) ion exchange preparation conditions on membrane properties," pp. 182–195, 2013, doi: 10.3390/membranes3030182

[19] J. Sharma and V. Kulshrestha, "Advancements in polyelectrolyte membrane designs for vanadium redox flow battery (VRFB)," *Results Chem.*, vol. 5, no. August 2022, p. 100892, 2023, doi: 10.1016/j.rechem.2023.100892

[20] S. Banerjee and K. K. Kar, "Impact of degree of sulfonation on microstructure, thermal, thermomechanical and physicochemical properties of sulfonated poly ether ether ketone," *Polymer (Guildf).*, vol. 109, pp. 176–186, 2017, doi: 10.1016/j.polymer.2016.12.030

[21] F. Lulay, C. Weidlich, M. Valtiner, and C. M. Pichler, "Membrane degradation in redox flow batteries," *Green Chem. Lett. Rev.*, vol. 16, no. 1, 2023, doi: 10.1080/17518253.2023.2274529

[22] A. Verma and R. Pitchumani, "Investigation of mechanical behavior of membrane in polymer electrolyte fuel cells subject to dynamic load changes," *J. Fuel Cell Sci. Technol.*, vol. 11, no. 3, pp. 1–9, 2014, doi: 10.1115/1.4026551

[23] J. Cao, Z. Yuan, X. Li, W. Xu, and H. Zhang, "Hydrophilic poly(vinylidene fluoride) porous membrane with well connected ion transport networks for vanadium flow battery," *J. Power Sources*, vol. 298, pp. 228–235, 2015, doi: 10.1016/j.jpowsour.2015.08.067

[24] W. J. Zou, H. Y. Jung, and S. Jung, "Modeling and performance optimization of vanadium redox flow batteries," *J. Energy Storage*, vol. 121, no. April, p. 116601, 2025, doi: 10.1016/j.est.2025.116601

[25] Y. Yang, Q. Wang, S. Xiong, and Z. Song. "Research progress on optimized membranes for vanadium redox flow batteries," *Inorg. Chem. Front.*, vol. 11, no. 14, pp. 4049–4079, 2024, doi: 10.1039/d4qi00520a

# 9 Membranes for Electrolysis and Water Splitting

## 9.1 INTRODUCTION

Membranes play a vital role in the operation of an electrolyzer, acting as selective barriers that allow ion transport while preventing gas crossover [1]. The type of membrane chosen depends on the electrolyzer type and the electrolyte environment. In AWE, which involves water electrolysis with alkaline properties, diaphragms are used as porous separators that conduct $OH^-$ in concentrated alkaline solutions, typically 5–7 M KOH [2]. Diaphragms are crucial for ionic transport and the separation of oxygen and hydrogen gases, as well as other gases [3]. Conversely, PEMWE systems typically employ Nafion, a PFSA membrane known for its exceptional proton ($H^+$) conductivity [4]. These systems can operate at comparatively high current densities, generally in the range of 1–3 A $cm^{-2}$, while producing hydrogen of exceptionally high purity [5]. This superior performance primarily arises from the membrane's outstanding proton selectivity and chemical stability under electrolysis conditions [6]. The PFSA membrane also offers high chemical stability in acidic conditions, a significant advantage over traditional alkaline systems [7]. The latest development is anion-exchange membrane water electrolysis (AEMWE), which combines features of alkaline and proton-exchange systems [8]. It uses solid anion-exchange membranes (AEM) functionalized with quaternary ammonium or phosphonium groups. These membranes facilitate anion movement with moderate ionic conductivity of about 0.1 S/cm at temperatures between 60 and 80°C [9]. However, first-generation AEMs have limited durability, often less than 1,000 hours, during which their performance declines significantly. Current research focuses on enhancing the chemical stability and ion-conducting capabilities of these membranes to make AEMWE a viable alternative to other technologies like AWE and PEMWE.

## 9.2 MEMBRANE TYPES AND MATERIALS

Electrolyzer performance is closely tied to the physical, chemical, and electrochemical properties of the membrane, as shown in Figure 9.1. Achieving ideal ion conductance is the main challenge, requiring the membrane to also act as a selective barrier to prevent gas crossover and to maintain mechanical and chemical integrity under harsh operational conditions [11]. Different types of AWE, PEMWE, and AEMWE require distinct membrane specifications in terms of both material composition and structural design.


DOI: 10.1201/9781003679363-9

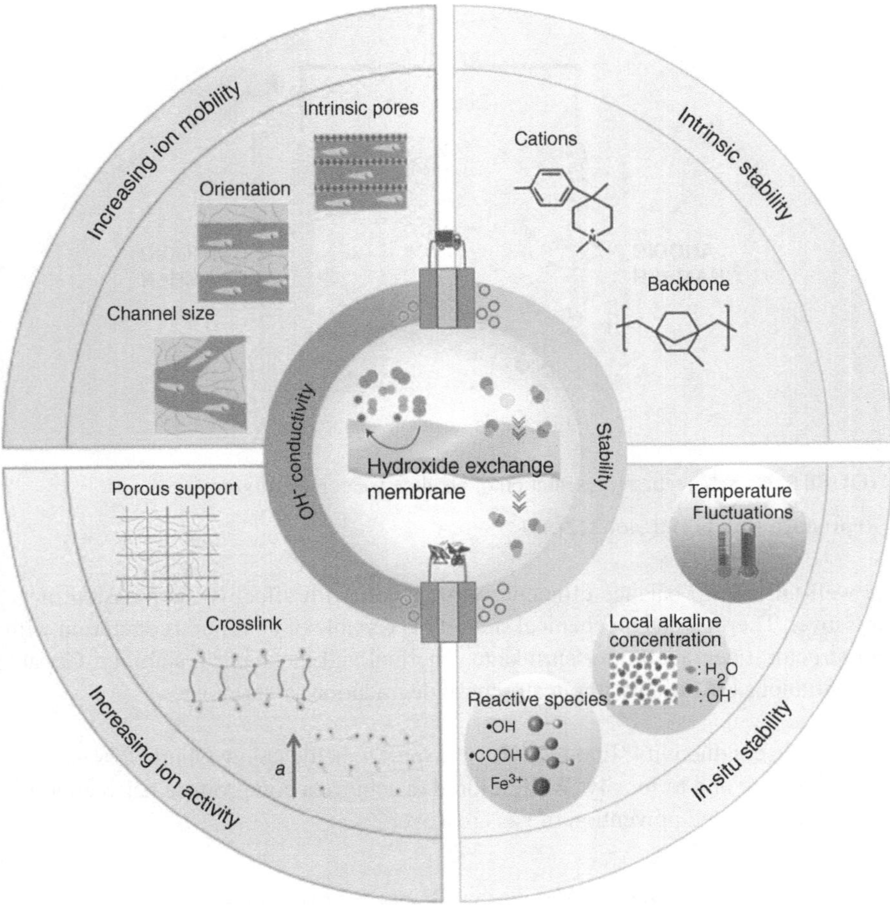


**FIGURE 9.1** Schematic illustration of the hydroxide exchange membranes towards water electrolysis.

**(Reproduced with permission [10].)**

### 9.2.1 Alkaline Water Electrolysis

For optimal OH- transfer, the diaphragm in an AWE should act as an electrical insulator to block current while allowing ion movement, as shown in Figure 9.2. AWE diaphragms must also possess high chemical stability to withstand damage from permeating ions in the highly alkaline environment of the electrodes, typically 5–7 M KOH [13]. Additionally, they need sufficient mechanical resistance to tolerate compression and ensure long life at elevated temperatures (>110°C). Commercial materials such as polymeric zirconium composite, specifically a polysulfone/zirconia composite characterized by a highly oxygenated, rigid structure, can meet these requirements. The AWE diaphragm must exhibit sufficient ionic conductivity with OH- transport to achieve low Ohmic losses. Porosity and tortuosity must

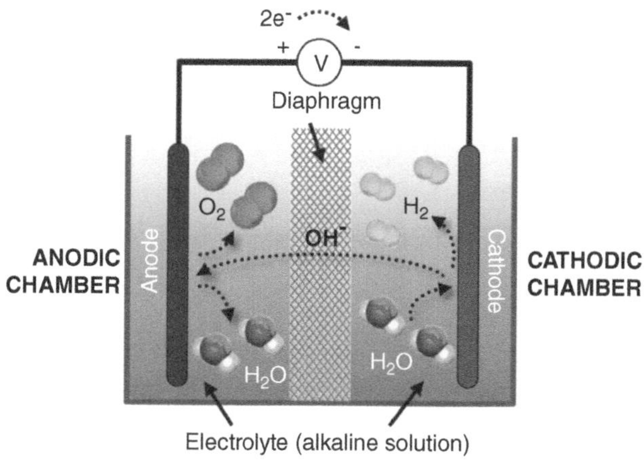


**FIGURE 9.2** Schematic representation of alkaline water electrolyzer.

**(Reproduce with permission [12].)**

be well-tailored to balance efficient ion transport with effective suppression of gas crossover. Thermal and mechanical stability is capable of continuous operation without structural degradation or shrinkage. Thermal and mechanical stability: Capable of continuous operation without structural degradation or shrinkage.

- Ionic conductivity: Enough OH- transport to achieve low Ohmic losses.
- Porosity and tortuosity: Well-tailored to achieve a compromise between ion transport and prevention of gas crossover.

### 9.2.2 Proton-Exchange Membrane Water Electrolyzers

The PFSA membranes, for example, Nafion, are considered the benchmark system in PEMWE because they offer the highest proton conductivity, around 0.1 S/cm, and maintain a highly stable, acidified environment. The membrane must fulfil several critical criteria. One of the main features of a proton exchange membrane (PEM) in a water PEM electrolyzer is that it acts as a physical barrier, preventing hydrogen from the cathode and oxygen from the anode from mixing. The membrane must have extremely high gas selectivity to minimize hydrogen and oxygen crossover because any mixing of these gases not only reduces the purity of the produced hydrogen but also poses significant safety risks due to the potential formation of explosive mixtures [14]. Effective gas separation enables the electrolyzer to produce high-purity hydrogen, which is especially important in applications requiring very high-quality hydrogen fuel, such as fuel cells and industrial hydrogen processes.

### 9.2.3 Anion-Exchange Membrane Water Electrolyzers

AEMs operate in a basic environment, so they require ionomers with chemically stable head groups that can endure these conditions. These ionomers are typically

**TABLE 9.1**
**Comparison of Electrolyzer Types and Membrane Roles**

| Parameter | AWE | PEMWE | AEMWE |
|---|---|---|---|
| **Electrolyte Medium** | 5–7 M KOH (liquid) | Solid acidic (PFSA) | Solid alkaline (functional polymer) |
| **Main Ion Transported** | $OH^-$ | $H^+$ | $OH^-$ |
| **Membrane Type** | Porous diaphragm (e.g., Zirfon) | PFSA membrane (e.g., Nafion) | Functionalized polymer (e.g., quaternary ammonium) |
| **Operating Temp. (°C)** | 60–110 | 50–80 | 50–80 |
| **Current Density (A cm$^{-2}$)** | 0.2–0.5 | 1–3 | 0.5–2 |
| **Gas Separation** | Moderate | Excellent | Moderate |
| **Hydrogen Purity** | ~99% | >99.9% | >99% |
| **Main Challenge** | Porosity control, carbonation | Cost, oxidative degradation | Alkaline stability, durability |

quaternary ammonium, imidazolium, or phosphonium. AEMWE are considered promising alternatives that integrate the advantages of both AWEs and PEMWEs into a single device. They offer benefits such as high operating current densities and pressures, Faradaic efficiencies exceeding 90%, and the use of non-precious metal catalysts with a pure water feed. However, hydrogen evolution is less favourable under alkaline conditions, as it is more efficient in acidic environments. Consequently, the overall voltage loss in AEMWEs arises mainly from two sources: ionic resistance within the membrane and activation losses associated with initiating the hydrogen and oxygen evolution reactions. In addition, interfacial electrical resistance between the membrane and catalyst layer contributes to the total Ohmic loss. Key membrane features are summarized in Table 9.1.

### 9.2.4 Hybrid and Composite Membranes

Hybrid and composite membranes have become some of the most advanced and promising alternatives to traditional polymer membranes, providing a balance of high ionic conductivity, mechanical strength, and long-term stability even in harsh electrochemical conditions (Table 9.2). Organic–inorganic composites combine polymers with ceramic or oxide fillers through physical or chemical interactions, effectively reducing swelling and gas crossover while stabilizing ion-conducting pathways. Reinforced membranes with porous supports like expanded polytetrafluoroethylene (ePTFE) or woven fabrics show excellent dimensional stability, allowing the use of thinner membranes without sacrificing mechanical strength or safety. Additionally, adding ionic liquids into polymer matrixes boosts ion mobility by creating stable, highly conductive regions, while also enhancing thermal and chemical resilience. The integration of advanced two-dimensional nanomaterials—such as MXenes and graphene oxide—brings ultrahigh surface area, tunable functional

**TABLE 9.2**
**Hybrid and Composite Membrane Strategies**

| Hybrid Type | Polymer Matrix | Inorganic Additive | Improvement Achieved |
|---|---|---|---|
| Polymer–Ceramic | PVDF, PEO | $Al_2O_3$, $ZrO_2$, $TiO_2$ | Mechanical strength, dimensional stability |
| Polymer–2D Material | PBI, PSU | Graphene oxide, MXene | Enhanced $OH^-$ pathways, barrier property |
| Reinforced Structure | PFSA, AEM | ePTFE, woven fabric | Reduced swelling, better durability |
| Polymer–Ionic Liquid | PVA, PEI | Imidazolium ILs | Improved conductivity, flexibility |

groups, and well-structured ion transport channels, thereby improving both ionic selectivity and conductivity. Overall, these innovations enable hybrid and composite membranes to surpass the traditional trade-offs between conductivity and durability, making them key materials for next-generation high-performance water electrolyzers [15].

## 9.3 ROLES AND KEY CHARACTERISTICS OF THE MEMBRANE

### 9.3.1 Chemical and Oxidative Stability

The PEM consistently operates under extremely harsh conditions, such as a strongly acidic environment and the presence of highly oxidative species, which result from the electrolysis process [16]. Reactive oxygen species and radicals generated at the anode interface may contact the membrane polymer backbone, thereby causing membrane thinning, the formation of pinholes, and even membrane rupture over extended operating periods [17]. Therefore, the membrane component must be highly resistant to chemicals and oxidation to withstand these tough conditions for thousands of hours without significant performance loss. This stability is essential for achieving long-term durability and ensuring the economic viability of PEMWE systems.

### 9.3.2 $OH^-$ Conductivity

Membranes in AEMWE must be $OH^-$ conductive, which is the fundamental requirement, as it directly influences the efficiency and the maximum achievable current density during operation. For AEMs to compete with PEMWE and AWE, they need to demonstrate high mobility for hydroxide ions, ideally exceeding 0.05 S/cm at the operating temperature. Achieving this involves designing ionomers with high ion-exchange capacity, substantial water absorption, and well-developed hydrophilic domains that facilitate efficient ion transport. However, unlike proton conduction in PFSA membranes, $OH^-$ conduction is inherently slower because hydroxide ions have lower mobility, which complicates materials design [18].

### 9.3.3 Alkaline Stability

Alkaline stability is a major obstacle to the large-scale implementation of AEMWE. To facilitate $OH^-$ transport, quaternary ammonium, imidazolium, and phosphonium groups are key candidates for cations. However, these groups tend to degrade under harsh alkaline conditions through mechanisms like nucleophilic substitution, Hofmann elimination, and oxidative pathways. Therefore, prolonged exposure to concentrated KOH or alkaline environments during electrolysis accelerates these reactions, leading to decreased ionic conductivity and membrane failure. As a result, researchers are exploring strategies such as protecting cationic groups with steric shields, adding fluorines to the backbone, and developing more stable head groups to extend membrane lifespan [19].

### 9.3.4 Membrane Thickness and Morphology Effects

Thickness and membrane morphology are key factors that determine the electrochemical performance, longevity, and efficiency of polymer electrolyte membranes, which are used in electrolyzers and fuel cells. Generally, a thinner membrane reduces ionic transport resistance, leading to higher conductivity and lower voltage losses; however, very thin membranes often lack mechanical stability, have high gas crossover, and lower operational durability. Conversely, thicker membranes provide greater mechanical strength and gas barrier properties, but increased thickness can still cause resistance issues that impact overall system efficiency. Additionally, membrane morphology, such as crystallinity, the degree of phase separation, pore structure, and channel orientation, significantly influences water transport, ion mobility, and mechanical robustness. For example, the highly structured phase-separated domains in Nafion-type PEMs create fully connected water channels for efficient proton conduction, while in AEMs, the distribution and stability of the polymer matrix's cationic groups determine hydroxide-ion transport pathways. In general, membrane thickness and morphology, shaped by advanced manufacturing techniques like electrospinning, phase inversion, or block copolymer design, enable precise balancing of conductance, selectivity, and durability. As a result, optimizing these parameters is crucial for creating membranes that achieve high ionic conductivity while maintaining long-term structural stability under harsh electrochemical conditions.

### 9.3.5 Membrane–Electrode Assembly Integration

Membrane–electrode assembly (MEA) integration is a crucial step in achieving efficient electrochemical performance and long-term stability of water electrolysis systems [20]. Properly engineered interfaces between membranes and electrodes are essential for eliminating interfacial resistance and enabling charge and mass transfer across the interface. In PEM electrolysis systems, strong interfacial interactions between perfluorosulfonic acid (PFSA) membranes and noble metal catalysts such as $IrO_2$ or Pt promote the formation of effective proton-conduction pathways. However, these interfaces are highly sensitive to local hydration conditions and catalyst dispersion, both of which significantly affect electrochemical performance. Similarly, in

AEM systems, the interaction between NiFeCo-based catalysts and the AEM occurs through hydroxide-ion transport. In such systems, interfacial compatibility and uniform distribution of the ionomer binder are critical for catalytic activity and durability. The ionomer binder within the electrode layers influences key aspects of the MEA, including catalyst utilization, triple-phase boundary development, and ionic conductivity. To improve these properties, various fabrication techniques, such as hot pressing, decal transfer, and in situ polymerization, are used to enhance interfacial contact between the membrane and catalyst layers. Each method presents different trade-offs regarding uniformity, mechanical durability, and scalability. The extent to which these integration strategies improve efficiency, durability, and long-term commercial viability remains a primary focus in advancing MEA design for electrolysis technologies.

## 9.4 PERFORMANCE LIMITATIONS IN ELECTROLYZERS

### 9.4.1 Mechanical Integrity Under Pressure

Most PEM water electrolyzers are designed to operate at high pressures to lower downstream hydrogen compression costs and improve system efficiency [21]. Under these conditions, the membrane faces significant mechanical stress caused by water uptake–induced swelling, pressure differences across the cell, and repeated thermal or hydration cycles [22]. If the membrane lacks adequate mechanical strength, it may deform, creep, or even rupture, which can lead to efficiency losses or complete system failure. Consequently, the membrane must be engineered to withstand dimensional changes while maintaining structural integrity under both static and dynamic stresses, ensuring reliable long-term performance during high-pressure operation.

### 9.4.2 Compatibility with Catalysts

The operation of a PEM water electrolyzer heavily depends on how well the membrane and the electrocatalyst layers are integrated. PFSA membranes like Nafion facilitate easy deposition and bonding of iridium-based catalysts at the anode, which are primarily responsible for the oxygen evolution reaction under acidic conditions. Strong compatibility between the membrane and catalyst not only ensures efficient proton transport across the interface but also enhances the stability of the catalyst layer, extending its lifespan by reducing delamination and performance loss during long-term operation. This interfacial synergy is a key factor behind the high current densities and effective electrochemical activity achieved by PEMWEs. Additionally, PEM membrane developers face challenges such as the high cost of precious metal catalysts, limited lifespans of around 5,000 hours, and degradation caused by local dehydration or over-oxidation. Researchers are exploring next-generation PFSA materials and reinforced composite membranes in an effort to improve quality and durability [23].

### 9.4.3 Mechanical Robustness

Mechanical robustness is essential because AEM undergo repeated cycles of hydration and dehydration during start-up, shutdown, and load fluctuations. Excessive water absorption, while generally beneficial for ion conductivity, can cause extensive swelling and compromise alkaline stability, which is a significant challenge for large-scale use of AEMWE. To improve $OH^-$ transport, quaternary ammonium, imidazolium, and phosphonium groups are key options for cations. However, these groups tend to degrade under harsh alkaline conditions through mechanisms such as nucleophilic substitution, Hofmann elimination, and oxidation. Prolonged exposure to concentrated KOH or alkaline environments during electrolysis accelerates these reactions, leading to reduced ionic conductivity and membrane failure. Therefore, researchers are exploring strategies like shielding cationic groups with steric barriers, adding fluorines to the backbone, and designing more stable head groups to extend membrane lifespan, maintain dimensional stability, and improve the separation of the membrane-electrode assembly (MEA). Mechanical stress is particularly intense in pressurized electrolysis systems, where membranes must withstand chemical attack and tensile forces. Methods to enhance the mechanical properties include using polymer backbones that are cross-linked, incorporating electrospun nanofibers as reinforcement matrixes, and creating hybrid organic–inorganic composites to balance water uptake with structural support.

### 9.4.4 Minimization of Carbonate Formation and Crossover

Mechanical robustness is crucial because AEM go through repeated cycles of hydration and dehydration during start-up, shutdown, and load changes, as shown in Figure 9.3. Excessive water absorption, although generally helpful for ion conductivity, can lead to swelling and weaken alkaline stability, which is a major challenge for large-scale use of AEMWE. To enhance $OH^-$ transport, quaternary ammonium, imidazolium, and phosphonium groups are key choices for cations. However, these groups often degrade under harsh alkaline conditions through mechanisms like nucleophilic substitution, Hofmann elimination, and oxidation. Prolonged exposure to concentrated KOH or alkaline environments during electrolysis speeds up these reactions, reducing ionic conductivity and causing membrane failure [25]. Therefore, researchers are exploring strategies such as shielding cationic groups with steric barriers, adding fluorines to the backbone, and designing more durable head groups to extend membrane lifespan, keep dimensional stability, and improve the separation of the MEA. Mechanical stress is especially high in pressurized electrolysis systems, where membranes need to resist chemical attack and tensile forces. Ways to improve mechanical properties include using cross-linked polymer backbones, adding electrospun nanofibers as reinforcement, and developing hybrid organic–inorganic composites to balance water uptake and structural support. The reduction of carbonate generation and crossover is one of the main difficulties of AEMWEs, especially those using ambient air. The $CO_2$ from the surrounding area dissolves in the electrolyte and reacts with $OH^-$ ions, producing carbonate ($CO_3{}^{2-}$) and bicarbonate ($HCO_3{}^-$) species. These anions have considerably lower mobility than $OH^-$; thus, ionic conductivity declines in the system and as a result, the efficiency of the system is lowered. Besides that, the carbonate crossover can contaminate the electrodes and accelerate the degradation of catalysts and electrodes. To address this challenge and ensure sustained operational stability, advanced strategies are being

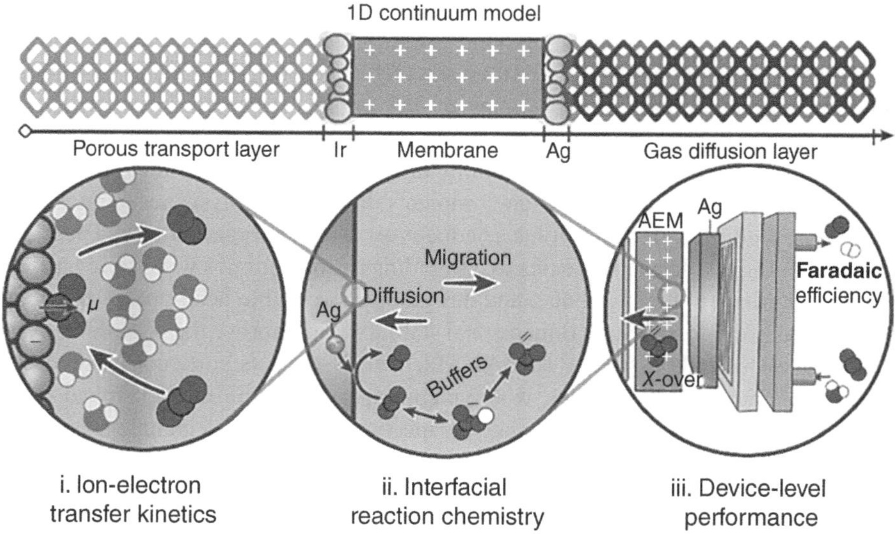


**FIGURE 9.3** Mechanism of $CO_2$ reduction and crossover in membrane electrode assemblies. **(Reproduce with permission [24].)**

explored, including the development of $CO_2$-tolerant AEMs, membranes engineered with selective anion transport channels, and the implementation of $CO_2$-free feedwater systems. The latest research in this field is mainly concerned with modifications of the polymer backbone, cross-linking properties, and hybrid composite membranes (incorporating inorganic fillers such as silica or zirconia) to improve not only conductivity but also their mechanical stability. The main challenge at the moment is the achievement of stability and durability (>1000h) under alkaline conditions, which is still the bottleneck before AEMWE can be considered as an alternative to commercial PEMWE and AWE technologies.

### 9.4.5 Membrane Degradation Mechanisms

One of the most critical issues that continues to pose a barrier to the longevity and cost-effectiveness of water electrolysis systems is membrane degradation. In PEMWE, PFSA membranes are continuously exposed to ROS like •OH and hydroperoxyl •OOH radicals, which cause chain scission, thinning, and reduced proton conductivity. Conversely, AEMs are impacted by nucleophilic attack on cationic head groups and Hofmann elimination of quaternary ammonium groups, leading to a loss of ionic functionality and mechanical embrittlement in alkaline conditions. Mechanical stresses such as repeated swelling and dehydration cycles induce fatigue, crack growth, and interfacial delamination, accelerating performance decay in both PEM and AEM systems. In AWE, diaphragms are particularly vulnerable to long-term exposure to concentrated KOH, resulting in carbonate formation, pore clogging, and eventual chemical degradation. Longevity assessments using accelerated stress tests like Fenton's test for oxidative stability in PEMs, combined with in situ monitoring of ionic conductivity, gas crossover, and dimensional stability, provide crucial insights into the failure mechanism. Case studies consistently demonstrate that degradation results from the interaction of chemical, mechanical, and electrochemical factors, underscoring the need for specialized treatments such as radical scavengers, chemically stable head groups, reinforced composites, and protective interlayers [26].

### 9.4.6 Water Management in Membranes

Efficient water management is essential for the performance and durability of membranes used in both PEMWE and AEMWE systems. The proton conductivity of PEM relies heavily on controlled hydration, as protons are transported via water channels within the polymer matrix. Insufficient hydration can cause a sharp decrease in conductivity and an increase in Ohmic resistance, while excessive water uptake may lead to swelling, mechanical weakness, and reduced gas barrier properties. Conversely, there is a strong correlation between membrane water content, hydroxide-ion conductivity, and distribution. However, if the membrane becomes too wet, the chemical degradation of functional groups accelerates, and the membrane surface can experience mechanical stress. Therefore, the membrane must maintain water balance while preserving dimensional stability throughout operation. Measures to improve water management include using hydrophilic or inorganic fillers to create more stable water pathways, designing membranes with cross-linked or more robust structures to prevent swelling, and employing ionomer chemistries that sustain consistent water content under varying operating conditions. Proper water management is one of the most effective approaches to achieve high ionic conductivity; it also minimizes gas crossover and prolongs membrane lifetime in electrolyzer environments.

### 9.4.7 Membrane Performance Metrics

Membrane performance metrics for an electrolyzer include electrochemical, mechanical, thermal, and chemical properties, each directly affecting the efficiency, lifespan, and safety of hydrogen production. A key parameter is ionic conductivity, which relates to the short-circuit current in PEMWE or hydroxide-ion transport in AWE and AEMWE systems. High ionic conductivity decreases internal Ohmic losses and boosts energy efficiency but must be balanced against mechanical and dimensional stability to prevent excessive swelling or deformation. For example, in PEMWE, proton conductivity is usually about 0.1 S/cm at 60–80°C, with hydration levels and the density of sulfonic acid groups in PFSA membranes being primary factors. AEMs exhibit OH- ion conductivity in the range of 0.05–0.1 S/cm, influenced mainly by water content, polymer flexibility, and the chemical stability of functional groups such as quaternary ammonium, imidazolium, or phosphonium. In AWE diaphragms, ions are transported through pores filled with electrolyte; conductivity depends on porosity, pore connectivity, and tortuosity. Gas crossover and selectivity are crucial parameters that determine a membrane's ability to prevent hydrogen and oxygen mixing. Excessive crossover reduces purity, increases safety risks, and may cause parasitic reactions. Dense perfluoro sulfonic acid (PFSA) membranes in PEM water effectively minimize gas crossover, even at high pressures. Conversely, AEMs and AWE diaphragms are more prone to gas diffusion, making factors like pore structure, cross-linking, and membrane thickness key design considerations.

PEMWE membranes typically require about 25–30 wt% water content to ensure high proton mobility. In AEMWE systems, excess water absorption can lead to carbonate formation from $CO_2$ ingress, causing short-term performance loss. AWE diaphragms rely on circulating alkaline electrolytes for hydration; however, too much electrolyte uptake can alter pore morphology and transport routes. Proper water management, through hydrophilic additives or engineered architectures, is essential across all systems. The mechanical and dimensional stability of membranes determines their ability to withstand compression, hydration/dehydration cycles, and thermal expansion during operation. PEMWE membranes must endure assembly pressures and repeated hydration without cracking or creeping. AEMs require chemically robust backbones and reinforcement to reduce swelling and degradation under alkaline conditions. AWE diaphragms must maintain pore integrity and structural strength under continuous electrolyte flow and pressure. Chemical and oxidative stability are equally important, as membranes operate in highly reactive environments. PEM membranes can be attacked by radicals generated during oxygen evolution, while AEMs are susceptible to nucleophilic degradation, Hofmann elimination, and carbonate formation. AWE diaphragms must keep their chemical composition stable under prolonged exposure to concentrated KOH. Accelerated ageing tests, exposing membranes to extreme pH, radicals, and heat, are often used to assess durability. Thermal stability ensures that conductivity and mechanical properties are maintained at operating temperatures—up to about 80°C for PEMWE and AEMWE, and above 110°C for AWE. Operating beyond these temperatures accelerates degradation, reduces conductivity, and weakens structural integrity [27].

These factors collectively determine the durability and operational lifetime of membranes: PEMWE membranes often last around 5,000 hours in lab conditions, while AEMWE membranes may degrade within about 1,000 hours. Although AWE diaphragms are chemically durable, they are subject to gradual mechanical wear or fouling, necessitating periodic replacement. Compatibility with both electrolytes and catalysts is another essential requirement. Membranes must maintain close interfacial contact with electrodes without causing delamination or chemical damage. PFSA membranes, for example, are compatible with noble metal catalysts like platinum and iridium oxide, while AEMs are designed for non-precious metal catalysts that enable hydroxide-ion conduction. In addition to these properties, several physical and electrochemical parameters, such as swelling ratio, ion-exchange capacity, porosity, tortuosity, and electrochemical stability window, further influence membrane performance and guide design improvements. In the end, creating next-generation membranes requires balancing ionic conductivity, gas selectivity, mechanical strength, chemical stability, and long-term durability. Hybrid and composite membrane approaches featuring polymer–ceramic structures, reinforced frameworks, and new functional groups are showing promise in improving these parameters simultaneously, opening the way for cost-effective, durable electrolyzers for industrial-scale hydrogen production.

## 9.5 FUTURE PERSPECTIVES ON ELECTROLYZERS

### 9.5.1 Recent Innovations and Advances

For many years, the membrane has been a limiting factor in the performance of water electrolysis devices operating in alkaline media. Recently, new high-performance ion-solvating membranes have emerged as a breakthrough technology, enabling hydroxide conductivities comparable to 1 M KOH benchmark electrolytes while functioning at lower alkalinity. These membranes, often based on polymers with specifically tailored free volume and polar functional groups, facilitate rapid $OH^-$ transport through Grotthuss-like and vehicular mechanisms. These improvements allow the use of less concentrated, less corrosive electrolytes, enhancing system safety and reducing balance-of-plant costs, while maintaining high current densities (>1 A/cm$^2$). From a future perspective, membranes and catalysts free of PGM are critical milestones for advancing AEMWE technology. An example worth highlighting is polysulfone-interpenetrating covalently cross-linked polybenzimidazole (C-PBI) with TPPR-12F structures, which has demonstrated the ability to sustain a current density of 500 mA/cm$^2$ for 160 hours under operating conditions far surpassing the durability of commercial FAA-3-50 membranes. The adoption of PGM-free catalysts paired with these advanced membranes offers a viable solution to the high system costs that hinder large-scale, sustainable hydrogen production. Additionally, polyphenylene-based AEMs are gaining attention due to their exceptional chemical and mechanical stability. These membranes, made from trifluoro phenyl (TFP) monomers, exhibit hydroxide conductivities exceeding 100 mS/cm at 80°C without losing structural integrity, even after harsh alkaline treatments. Long-term tests have shown stable operation beyond 1,000 hours, confirming their capability for continuous green hydrogen production in industrial settings. Furthermore, their solid aromatic backbone and resistance to nucleophilic attack help prevent degradation mechanisms typically caused by quaternary ammonium-functionalized membranes. Collectively, these innovations represent a significant advancement in the evolution of AEMWE technology but also bring the industry closer to pilot-scale deployment. By addressing fundamental challenges such as alkali stability, conductivity at lower KOH concentrations, and cost reduction through PGM-free integration, these membrane systems are now positioned to enable scalable, durable, and affordable hydrogen production.

### 9.5.2 New Material Strategies in Electrolyzers

Composite membranes are increasingly developed by incorporating inorganic nanoparticles or ceramic fillers to enhance mechanical strength, thermal stability, and ionic transport. Reinforced membrane structures employ porous supports or cross-linked polymer networks to effectively control swelling and dimensional variations during operation. Functional group engineering involves modifying membrane chemistry to improve ion selectivity, reduce gas crossover, and increase resistance to chemical degradation. In addition, membranes designed for high-temperature operation enable faster reaction kinetics and improved electrolyzer efficiency by maintaining stable performance at elevated temperatures. Electrolysis membrane design is essentially a balance among the properties of ion conductivity, chemical and mechanical stability, gas selectivity, and operational durability. Breakthroughs in material chemistry and membrane structures are required to enable commercial implementation and long-term efficiency of all three electrolyzer technologies.

## 9.6 CONCLUSION

The rapid progress in membrane science within alkaline water electrolysis (AWE) highlights the importance of material innovation for large-scale, low-cost green hydrogen production. Ion-solvating membranes, which can achieve ionic conductivities like those of concentrated alkaline electrolytes while functioning under milder conditions, have effectively addressed major challenges related to durability and safety in electrolysis systems. Moreover, polyphenylene-based backbones functionalized with TFP groups demonstrate excellent hydroxide-ion conductivity and exceptional chemical resistance for long-term operation, marking a significant advancement towards industrial reliability. Together, these breakthroughs shift the focus from incremental improvements to advanced design strategies that involve engineering polymers to enhance hydroxide-ion transport, using ion-solvating or interpenetrating networks to boost dimensional stability, and developing membrane–catalyst architectures to maximize performance while minimizing dependence on PGMs. As AEMWE approaches commercialization, future research must aim to translate these laboratory innovations into scalable, mechanically durable, and economically viable membrane platforms capable of supporting high current densities under industrial conditions.

## REFERENCES

[1] M. F. Ahmad Kamaroddin et al., "Membrane-based electrolysis for hydrogen production: A review," *Membranes (Basel)*, vol. 11, no. 11, pp. 1–27, 2021, doi: 10.3390/membranes11110810

[2] D. Henkensmeier et al., "Separators and membranes for advanced alkaline water electrolysis," *Chem. Rev.*, vol. 124, no. 10, pp. 6393–6443, 2024, doi: 10.1021/acs.chemrev.3c00694

[3] A. Abdel Haleem et al., "Innovative membrane with selective gas permeability for alkaline water electrolysis: Dependable cell performance under industrial conditions," *J. Power Sources*, vol. 587, no. May, p. 233709, 2023, doi: 10.1016/j.jpowsour.2023.233709

[4] A. Ghosh, T. Sharma, and J. Pandey, "Review on proton conducting membrane for PEM water electrolyser: A sustainable approach for green hydrogen production," *J. Electroanal. Chem.*, vol. 993, no. June, p. 119286, 2025, doi: 10.1016/j.jelechem.2025.119286

[5] G. H. A. Wijaya, K. S. Im, and S. Y. Nam, "Advancements in commercial anion exchange membranes: A review of membrane properties in water electrolysis applications," *Desalin. Water Treat.*, vol. 320, no. May, p. 100605, 2024, doi: 10.1016/j.dwt.2024.100605

[6] P. P. Shwe Sin, S. H. Wai, Y. Ota, K. Nishioka, and Y. Suzuki, "Performance recovery of proton exchange membrane electrolyzer degraded by metal cations contamination," *Int. J. Hydrogen Energy*, vol. 53, no. September 2023, pp. 86–92, 2024, doi: 10.1016/j.ijhydene.2023.12.049

[7] H. Nguyen et al., "Hydrocarbon-based Pemion™ proton exchange membrane fuel cells with state-of-the-art performance," *Sustain. Energy Fuels*, vol. 5, no. 14, pp. 3687–3699, 2021, doi: 10.1039/d1se00556a

[8] M. M. Shanbhag, S. Mishra, N. P. Shetti, T. Palanivel, S. S. Kalanur, and B. G. Pollet, "Advancements and challenges in anion exchange membrane technology for sustainable energy applications," *Mater. Today Energy*, vol. 53, no. July, 2025, doi: 10.1016/j.mtener.2025.101976

[9] Z. Feng, G. Gupta, and M. Mamlouk, "A review of anion exchange membranes prepared via Friedel-Crafts reaction for fuel cell and water electrolysis," *Int. J. Hydrogen Energy*, vol. 48, no. 66, pp. 25830–25858, 2023, doi: 10.1016/j.ijhydene.2023.03.299

[10] W. Wang, Y. Sun, X. Liu, M. Wei, C. Zhao, K. G. Zhou, and M. Wu, "Hydroxide exchange membranes towards water electrolysis and fuel cells: A review on the recent advances, challenges and opportunities," *Adv. Membr.*, vol. 4. p. 100111, 1 Jan 2024, doi: 10.1016/j.enchem.2022.100087

[11] E. E. Nzaba Madila, A. Makhsoos, M. M. Shanbhag, and B. G. Pollet, "Advancements in electrolyser stack performance: A comprehensive review of Latest technologies and efficiency strategies," *Int. J. Hydrogen Energy*, vol. 144, no. July 2024, pp. 1168–1189, 2025, doi: 10.1016/j.ijhydene.2025.04.047

[12] E. Amores, M. Sánchez, N. Rojas, and M. Sánchez-Molina, "Renewable hydrogen production by water electrolysis," in *Sustainable Fuel Technologies Handbook*, pp. 271–313, Academic Press, 2021.

[13] K. S. Im et al., "Development and characterization of engineering plastic diaphragm for alkaline water electrolysis," *Desalin. Water Treat.*, vol. 320, no. August, p. 100692, 2024, doi: 10.1016/j.dwt.2024.100692

[14] M. Schalenbach and D. Stolten, "High-pressure water electrolysis: Electrochemical mitigation of product gas crossover," *Electrochim. Acta*, vol. 156, pp. 321–327, 2015, doi: 10.1016/j.electacta.2015.01.010

[15] M. Ö. Seydibeyoğlu, A. Dogru, J. Wang, M. Rencheck, Y. Han, L. Wang, et al., "Review on hybrid reinforced polymer matrix composites with nanocellulose, nanomaterials, and other fibers," *Polymers*, vol. 15, no. 4, p. 984, 2023, doi: 10.3390/polym15040984

[16] S. Shiva Kumar and V. Himabindu, "Hydrogen production by PEM water electrolysis – A review," *Mater. Sci. Energy Technol.*, vol. 2, no. 3, pp. 442–454, 2019, doi: 10.1016/j.mset.2019.03.002

[17] E. Wallnöfer-Ogris, F. Poimer, R. Köll, M. G. Macherhammer, and A. Trattner, "Main degradation mechanisms of polymer electrolyte membrane fuel cell stacks – Mechanisms, influencing factors, consequences, and mitigation strategies," *Int. J. Hydrogen Energy*, vol. 50, pp. 1159–1182, 2024, doi: 10.1016/j.ijhydene.2023.06.215

[18] L. Liu, H. Ma, M. Khan, and B. S. Hsiao, "Recent advances and challenges in anion exchange membranes development/application for water electrolysis: A review," *Membranes (Basel)*, vol. 14, no. 4, 2024, doi: 10.3390/membranes14040085

[19] Z. Tang, B. Wu, K. Yan, J. Luo, M. U. Haq, and L. Zeng, "Long-term stability for anion exchange membrane water electrolysis: Recent development and future perspectives," *Futur. Batter.*, vol. 5, no. January, p. 100024, 2025, doi: 10.1016/j.fub.2025.100024

[20] K. Lawand, S. Nuggehalli Sampathkumar, Z. Mury, and J. Van Herle, "Membrane electrode assembly simulation of anion exchange membrane water electrolysis," *J. Power Sources*, vol. 595, no. January, p. 234047, 2024, doi: 10.1016/j.jpowsour.2023.234047

[21] C. R. Wang et al., "Proton Exchange Membrane (PEM) water electrolysis: Cell-level considerations for gigawatt-scale deployment," 2025, doi: 10.1021/acs.chemrev.3c00904

[22] E. Urbano, E. Pahon, N. Yousfi-Steiner, and M. Guillou, "Accelerated stress testing in proton exchange membrane water electrolysis - critical review," *J. Power Sources*, vol. 623, no. September, p. 235451, 2024, doi: 10.1016/j.jpowsour.2024.235451

[23] N. Sezer, S. Bayhan, U. Fesli, and A. Sanfilippo, "A comprehensive review of the state-of-the-art of proton exchange membrane water electrolysis," *Mater. Sci. Energy Technol.*, vol. 8, no. July 2024, pp. 44–65, 2025, doi: 10.1016/j.mset.2024.07.006

[24] E. W. Lees, J. C. Bui, O. Romiluyi, A. T. Bell, and A.Z. Weber, "Exploring CO2 reduction and crossover in membrane electrode assemblies," *Nat. Chem. Eng.*, vol. 5, pp. 340–353, 1 May 2024.

[25] H. Becker, J. Murawski, D. V. Shinde, I. E. L. Stephens, G. Hinds, and G. Smith, "Impact of impurities on water electrolysis: a review," *Sustain. Energy Fuels*, vol. 7, no. 7, pp. 1565–1603, 2023, doi: 10.1039/d2se01517j

[26] N. A. Amin, N. Abdullah, and N. A. Karim, "Support material design strategies for oxygen evolution reaction with degradation resistance and functional enhancement," *Int. J. Hydrogen Energy*, vol. 206, p. 153346, 4 Feb. 2026, doi: 10.1016/j.ijhydene.2025.153346

[27] M. Bonanno et al., "Review and prospects of PEM water electrolysis at elevated temperature operation," *Adv. Mater. Technol.*, vol. 9, no. 2, pp. 1–25, 2024, doi: 10.1002/admt.202300281

# 10 Industrial and Sustainable Perspectives on Membrane Technologies

## 10.1 INTRODUCTION

Energy is a global challenge that requires a sustainable and resilient future. Growing concerns over climate change, resource depletion, and population expansion have led to a search for future energy solutions, a major area of innovation. So, improving the energy efficiency through new infrastructure can enhance the positive global transformation. Polymeric membrane-based electrochemical energy storage systems represent a promising approach in reducing electricity demand through efficient energy conversion, storage, and decentralized power management. Although industrial-scale implementation of the membrane requires evaluation that extends beyond simple performance, along with its scalability, reproducibility, cost-effectiveness, and overall sustainability. The industrial viability of the membrane depends on the key performance metrics such as conductivity, selectivity, and durability. Whereas membranes intended for redox flow batteries, supercapacitors, fuel cells, and emerging sodium-ion or lithium-ion based batteries need to be produced in uniform, defect-free batches to ensure reliable performance at scale-level production [1]. In addition, the fabricated membranes must be compatible with automated roll-to-roll manufacturing to support continuous large-scale production, along with the environmental aspect of minimizing the use of hazardous solvents and energy-intensive processing that integrate eco-friendly materials and related green manufacturing strategies, which ensure sustainability and circular economy principles. Typical fluorinated ionomers like perfluorosulfonic acids (PFSAs) are major contributors to the carbon footprint because of their long-lasting nature in the environment (PFAS problem), highly energy-demanding synthesis, and a few recycling routes [2]. Hydrocarbon alternatives, bio-based polymers, and green processing methods such as solventless extrusion, aqueous casting, or additive manufacturing are the areas that researchers are actively working on for the purpose of reducing emissions related to the whole product life cycle and the reliance on rather harmful chemicals [3]. To address end-of-life concerns in electrochemical storage devices, the application of life cycle assessment evaluates energy return time, carbon footprint, and circular economy potential of membrane materials utilized in electrochemical devices. LCA offers a comprehensive approach that allows both environmental and economic trade-offs to be evaluated at the same

DOI: 10.1201/9781003679363-10

time, and also avoids the limitations associated with evaluating these factors independently. Thereby providing an essential tool for understanding the complementary roles, long-term sustainability, and practical feasibility of membrane-based electrochemical technologies.

## 10.2 COMMERCIALIZATION AND PRODUCTION

Applications of membranes in electrochemical storage devices have undergone a significant transformation from laboratory scale to industrial deployment. Among them, Ion exchange membrane technologies are experiencing substantial prominence within the energy sector, particularly in carbon capture, chemical production, utilization applications, and electrochemical energy storage, which demand innovation to achieve net-zero emission targets. The gaining attention is due to the high conductivity and desired selectivity [4]. The successful commercialization of these technologies depends on improving their techno-economic viability, which can be achieved by enhancing the performance of IEM technologies. The choice of polymers is a key factor in the commercialization of the membrane. It is desirable that the polymer selected is flexible and has high mechanical strength along with chemical stability. To achieve product quality, cost-effectiveness, and commercial viability of the product, it is imperative to optimize the polymerization process as well as the scale-up process.

There are a number of factors that need to be taken into consideration during the scale-up of the membrane. These include reactor size, availability of the feedstock materials, and the processing steps. It is also important to include a pilot plant during the scale-up process. This is because the pilot plant helps in the evaluation of the process [5]. Owing to that, the production of hydrogen through proton exchange membrane and anion-exchange membrane electrolyzers is crucial to industrial decarbonization and advancement of hydrogen-based transportation [6]. Hydrogen production through electrolyzers is growing quickly in the market. It is expected to increase nearly sixfold by 2030 compared to 2022. This highlights its rising role in the shift to sustainable energy. Similarly, redox flow batteries, particularly vanadium flow batteries, are expected to experience significant market expansion, with an annual growth of approximately 26.5% through 2020, demonstrating their industrial feasibility for large-scale energy applications. A comparative overview of these industrially relevant membranes, including their key characteristics and applications, is summarized in Table 10.1. Perfluorinated cation-exchange membranes were often utilized in fuel cells. Notably, Nafion membrane, composed of polytetrafluoroethylene with perfluoropolyether side chains and sulfonate groups at the end, increases the chemical stability of the membrane. The economic viability of IEM production mainly depends on future improvements and lower material costs. Recent techno-economic studies and process analyses offer useful insights into the practicality, challenges, and potential methods to enhance the production of IEM technologies used in energy storage and chemical manufacturing. Collectively, these trends emphasize the expanding commercialization and industrial relevance of membrane technologies in enabling the global transition towards sustainable energy systems [7].

**TABLE 10.1**
**Comparative Overview of Industrially Relevant Membranes**

| Membrane Type | Base Polymer | Proton Conductivity (s $cm^{-1}$) | Thermal Stability (°C) | Key Limitations | Typical Electrochemical Device |
|---|---|---|---|---|---|
| **Nafion** | PFSA | 0.10–0.20 | 180–200 | High cost; PFAS environmental concern | PEMFC, PEMWE |
| **SPEEK** | Sulfonated PEEK | 0.05–0.10 | 160–180 | Chemical degradation under high humidity | PEMFC, VRFB |
| **SPI** | Sulfonated Polyimide | 0.03–0.08 | 180–220 | Brittle; limited water uptake | Fuel cells, electrolysis |
| **Hydrocarbon Composite** | SPEEK + $SiO_2$/ZIF-8 | 0.07–0.12 | 200–250 | Complex synthesis | PEMFC, RFB |
| **Biopolymer Based** | Cellulose, chitosan, alginate | 0.01–0.05 | 120–160 | Low durability; humidity sensitivity | Water electrolysis, sensors |

## 10.3 COST CONSIDERATIONS

To increase the global production of ion exchange membranes, their design must become more efficient and less expensive for flow battery and other electrolyzer applications. Currently, IEMs account for 35%–60% total stack costs, which itself represents about 40% capital expenditure needed to deploy all reviewed membranes. This means IEMs materials make up 14%–28% of total capital expenditure. High capital costs discourage early adoption and increase investment risk, even in well-established technologies [7]. The IEM market will grow 11% annual rate each year until 2028. This shows that the electrochemical energy sector is expanding. However, this forecast does not fully account for economic competitiveness. Many electrochemical systems using IEMs still struggle to meet cost benchmarks set by agencies like the U.S. Department of Energy, including a target to produce hydrogen at $1 per kilogram by 2030. This gap between technological progress and cost goals highlights the role of membranes in overall system economics [8]. Since IEMs significantly impact stack and capital costs, improving their affordability, durability, and performance is crucial to reducing overall ownership costs. Understanding membrane costs depends on raw materials, fabrication methods, lifespan, and replacement frequency. A recent DOE survey revealed that conservative investment strategies by utilities hinder the growth of commercial redox flow battery companies, despite the technology's economic viability. These strategies affect membrane costs or limit price increases for new materials that could pose investment risks. Furthermore, all commercially available membranes, such as Nafion® developed by DuPont, are based on perfluorinated polymer backbones with sulfonic acid side chains, which offer high mechanical strength, chemical stability, and proton conductivity, along with increased acidity and resilience. The cost of Nafion® membranes, around US$500 per square metre, accounts for up to 50% of the capital costs of flow battery stacks, impacting large-scale applications. Additionally, concerns are rising about the environmental impacts of manufacturing Nafion® and its by-products, perfluoroalkyl and polyfluoroalkyl substances (PFASs), which can contaminate drinking water and pose long-term health risks [9]. Achieving terawatt-scale renewable energy conversion and storage will require millions of square metres of ion exchange membranes. Due to cost and environmental considerations, replacing Nafion® membranes with low-cost, sustainable alternatives is necessary. Non-fluorinated hydrocarbon ion exchange membranes offer notable advantages over PFSA membranes in terms of performance, cost, and environmental impact. Over recent decades, extensive research has been carried out on fabricating hydrocarbon-based ion exchange membranes, including cation- and anion-exchange membranes for flow batteries and fuel cells. However, the stability of hydrocarbon-based anion-exchange membranes largely depends on the properties of their cationic groups. Many studies have shown that hydroxide ions can more easily degrade the polymer backbone, affecting the long-term stability and operational performance of the membrane. Moreover, large-scale validation of low-cost hydrocarbon membranes and their integration into commercial-scale flow battery systems with advanced chemistries has not yet been reported [10].

## 10.4 ENVIRONMENTAL IMPACT AND LCA

Electrochemical membranes offer promising opportunities that improve the sustainability of energy and chemical processes. Their high selectivity, efficient separation, and conversion, which lowers material waste and improves overall resource utilization. These systems consume less energy than conventional separation technologies [11]. Unlike thermally driven or high-pressure processes, electrochemical membranes operate under lower temperature and pressure conditions. As a result, overall energy consumption decreased, which directly reduced greenhouse gas emissions. In addition, these technologies operate based on electricity, and they can be easily coupled with renewable energy sources like wind power and solar, which enhances environmental performance [12]. Existing research on batteries has outlined several promising directions and methodological approaches. However, most impact assessment studies tend to focus on a limited range of environmental indicators, primarily cumulative energy demand and climate change, while overlooking other potentially significant impact categories. This limited study increases the risk of shifting environmental problems from the assessed environmental issues to areas that are not covered. Membrane-based electrochemical systems and life cycle environmental assessments have revealed substantial life cycle impacts, especially during the stages of material synthesis and fabrication, as summarized in Table 10.2. The membranes used in electrochemical storage devices are still in development compared with conventional technology and have various operational problems associated with their use in terms of cost and environmental impact. These membranes may have problems with electrode and membrane degradation under an electric field, problems with pH stability, and high electrode and ion exchange material costs, which may limit their long-term stability and scalability. Life Cycle Assessment has been used to quantify the environmental performance of batteries and other storage devices. This

**TABLE 10.2**
**Environmental Metrics and Lifecycle Considerations**

| Lifecycle Stage | Conventional PFSA Systems | Sustainable Alternatives | Mitigation Strategies |
|---|---|---|---|
| **Raw Material Extraction** | Fluorinated monomers from petrochemical feedstocks; high GHG emissions | Bio-based monomers (cellulose, lignin) | Feedstock diversification; renewable sourcing |
| **Membrane Fabrication** | Solvent-intensive (NMP, DMF); VOC release | Aqueous or solvent-free processing | Green solvent substitution; closed-loop recovery |
| **Operation** | Stable, high-performance | Comparable if optimized; may degrade faster | Surface functionalization; hybrid fillers |
| **End-of-Life** | Non-degradable; PFAS persistence | Reprocessable and biodegradable | Mechanical recycling; chemical upcycling |
| **LCA Carbon Footprint** | ~60–100 kg $CO_2$ eq $m^{-2}$ | ≤ 20 kg $CO_2$ eq $m^{-2}$ | Renewable energy integration; modular design |

framework is adopted because it provides broad coverage of the life cycle stages of the studied products, which quantifies their impacts and damage across environmental factors. In this context, techno-economic analysis provides insight into the cost-effectiveness and profitability of EMTs operational performance, evaluating factors such as operating costs, capital investment, and revenue potential. On the contrary, LCA assesses the environmental impacts of all of these technologies throughout their entire lifecycle, from the extraction of raw materials up to their disposal. Some of the significant parameters that are taken into consideration include energy consumption, water usage, GHG emissions, and ecological risks. Thus, it is very useful in helping policymakers, researchers, and stakeholders make better decisions that take into account both the economic and environmental benefits of the technologies [13].

### 10.4.1 Environmental and Social Impacts of Raw Materials

Raw materials are crucial in the development of energy storage devices. However, the use of these raw materials is a major challenge from an environmental and social perspective. There are major environmental effects associated with the mining and refining of crucial raw materials, by-products, and technology. Therefore, the environmental hazards and effects associated with resources used for raw materials are always a contentious issue. The significance of these sectors in attaining a low-carbon and carbon-neutral economy has generated the need to use raw materials that increase simultaneously with the development of these crucial sectors. The environmental costs associated with mining, quarrying, harvesting, and refining crucial raw materials for storage devices vary according to the particularity of the material and technology used, hence creating different environmental impacts. For instance, it has been postulated that the environment could become unstable due to the infrastructure that will be put in place for the mining and processing of metal ores. Mining and processing activities related to the use of metal compounds could be detrimental. There are other significant raw materials that could have their environmental hotspots related to various stages of their production processes. For instance, it is interesting to note that the main raw materials used in lithium-ion batteries include silicon, cobalt, and natural graphite. On the other hand, the basic raw materials that are used in nickel metal hydride batteries include cerium, lanthanum, neodymium, and praseodymium [14]. The mining and processing of cobalt and other rare earth compounds have medium to high potential environmental risk, while graphite has negligible risk. The greatest impact of a battery's environment life cycle is experienced during the extraction and processing of electrode materials. Cobalt is commonly used in conjunction with other metals such as aluminium, nickel, manganese, and lithium in the cathode of a lithium-ion battery [10]. Although battery cells and aluminium, the active cathode material, production has the greatest environmental impact to date, its impact on battery power is probably negligible. The environmental impact of battery power is mainly caused by essential raw materials such as cobalt. Poor labour practices, resource conflicts, health issues, and dislocation of communities are some of the social impacts of raw materials. The environmental impacts of raw materials include soil degradation, air pollution, water pollution, and habitat destruction. These are some of the interrelated problems that bring into focus the broader consequences of the increased global need for essential raw materials, which are necessary for contemporary manufacturing, information technology, and green energy. Social impact assessment, abbreviated as SIA, is a method of analysing, understanding, and evaluating the significance of the social impacts of a given product or service throughout its lifecycle. It assesses the importance of possible social impact and recommends strategies to avoid or minimize any adverse social impact while enhancing positive social impact.

## 10.5 STRATEGIES FOR SUSTAINABILITY: GREEN MATERIALS AND RECYCLING

High-performance materials are synthesized from green resources, and they have gained attention due to their low-cost and sustainable properties. Numerous green-derived sources are reliable resources for the preparation of multifunctional materials. Furthermore, the intrinsic structures of these green materials and functional materials can be tailored. In particular, rechargeable batteries such as lithium-ion batteries, supercapacitors, and other electrolyzers are considered the most efficient storage devices developed to date. However, these technologies have several limitations, high cost, low energy density, and safety concerns, which hinder their widespread application. To address issues, a notable innovation in the design of materials used in separators to improve performance, reduce the cost of rechargeable batteries, and electronic devices. Employing appropriate surface modifications and chemical functionalization of these biopolymers can tailor them to desirable properties. In parallel, solvent engineering plays an important role in sustainable membranes. The use of eco-friendly green solvents such as γ-valerolactone, Cyrene, and water significantly reduces volatile organic compound emissions. By preserving the economic value of materials and resources, the circular economy integrates sustainability. The EU list of Critical Raw Materials (CRMs) was introduced under the EU Raw Materials Initiative (RMI), and the European Commission has established a regulatory framework to promote sustainability and circularity in the electromobility battery value chain. This is in the quest to promote the use of batteries in various applications while ensuring that they follow sustainable practices. In the energy sector, the re-functionalization and re-casting of old PEM fuel cell membranes have the potential to regenerate their performance in the reduction of dependence on new materials of the PFSA type.

Requirements for the use of recycled materials, along with the measures aimed at improving the collection of batteries and their recycling rates, promote the recovery of valuable resources. In addition to this, non-rechargeable batteries should be considered, and it is essential to establish standards that promote battery sustainability and transparency. These standards should take into account the factors such as carbon footprint associated with the battery production, the security of the raw materials supply, and the ease with which batteries can be recycled, reused, or repurposed at the end of their life. To encourage more economic models, the commission also suggests revising the regulations for end-of-life vehicles by tying design considerations to end-of-life treatments, taking into account regulations that require recycled content for specific materials and components, which enhances recycling efficiency. This includes adopting product-as-service methods to reduce consumption of new materials, promoting use of sustainable alternatives by optimizing infrastructure, and reducing waste and pollution. These approaches together make the system more efficient and environmentally sustainable.

## 10.6 FEASIBILITY FOR INDUSTRIAL DEPLOYMENT

The industrial feasibility of sustainable membranes mainly depends on their successful production at a commercial scale. However, this scale-up process is not always straightforward, as it involves various challenges such as technical limitations, regulatory compliance, and economic considerations. Resolving these challenges is necessary for the effective large-scale application of membrane technologies in industry.

### 10.6.1 Technical Feasibility

A key challenge in the industrial deployment of sustainable membranes is that their performance does not match that of PFSA membranes. To meet similar performance standards, the membrane should undergo modifications to the membrane structure and material composition of hydrocarbon ionomers, biopolymer-based films, and composite membranes should ideally reach ionic conductivities of about $10^{-2}$ S/cm for proton exchange systems while maintaining structural and electrochemical durability over 5,000–20,000 hours of operation to compete with PFSA membranes[15]. In addition to conductivity, several other properties such as stronger mechanical stability, chemical resistance under extreme operating conditions, and high ion selectivity to function effectively over long cycles influence practical applications. Maintaining precise control over the thickness of the membrane in the range of (10–30 μm) remains difficult. Subsequently, the drying of the film and ensuring adequate porosity are also challenging because conventional drying methods often require high energy consumption. Another major issue is the integration between the membrane and the electrode. The interface between these components is not well matched to the internal resistance of the system, and can increase, which may reduce the overall efficiency of the device by 20%, so these technical challenges may influence the industrial deployment of membranes.

### 10.6.2 Regulatory and Policy Drivers

In terms of regulatory frameworks, government policies are increasingly driving changes in the membrane market. For instance, the EU's REACH regulation severely prohibits N-methyl-2-pyrrolidone, N, N-dimethylacetamide, and N, N-dimethylformamide solvents that are used during the fabrication of the membrane. As a result, there is a growing interest in developing water-based or non-volatile organic solvent systems, yet there remains a conflict. Another major issue is the increasing global concern regarding PFSA membranes such as Nafion®. Regulatory authorities of the EU and the United States are demanding fluorine-free membranes, such as sulfonated hydrocarbon membranes. At the same time, climate-focused policy measures such as carbon taxes and clean technology commitments are encouraging industries to shift towards more sustainable and low-carbon materials. Thus, meeting regulatory requirements is no longer only about safety and compliance; it can also provide companies with a competitive advantage in the evolving membrane market.

### 10.6.3 Economic Drivers

The economic feasibility of membrane technologies is still the main reason that these technologies are hardly utilized on a large scale. The price of PFSA membranes varies from \$600 to \$1200 per square metre, which makes 40% of the total cost of a fuel cell or a redox flow battery stack. By contrast, the production of hydrocarbon membranes, such as SPEEK, in terms of costs can be close to \$40–60 per square metre. However, at the system level, electrolyte costs typically account for 43–55% of the total expenses, exceeding membrane contributions. The newly emerging aqueous organic flow batteries deal with the problem of high electrolyte costs, mainly because of the lack of stability of organic active materials. As a result, it became difficult to achieve the \$100/kWh target for energy storage. Reports from the U.S. Department of Energy also suggest that current performance levels are not yet sufficient to satisfy large-scale manufacturing that would enable economies of scale. Overall cost of membrane, conductivity, selectivity, durability, and their impact on energy efficiency remain key economic factors shaping the commercial future of electrochemical storage technologies.

### 10.6.4 Social and Supply Chain Considerations

Block design is more flexible because of the supply chain and social factors. It is expected that the continued need for stationary energy storage will require the extraction of essential raw materials that include nickel, cobalt, and lithium. However, the challenges that concern labour conditions and welfare, and resource allocation in the mining regions were raised by the raw materials. The placement of the lithium-ion battery storage system, its design, and risks influence its public acceptance. Public acceptance assessment could be influenced by the installation of mid-to-large-sized stationary battery storage systems. This emphasizes the importance of social acceptance in the successful integration of battery technology. Raw material extraction in indigenous regions creates community and economic benefits.

## 10.7 INDUSTRIAL USE CASES AND SECTOR IMPACT

The industrial deployment of sustainable membrane technologies requires navigating multiple, often competing, scenarios as given in Figure 10.1. Water and energy are two essential building blocks that support both society and economic development. In recent years, membrane engineering has gained significant attention because of its crucial role in addressing global challenges such as water scarcity, climate change,

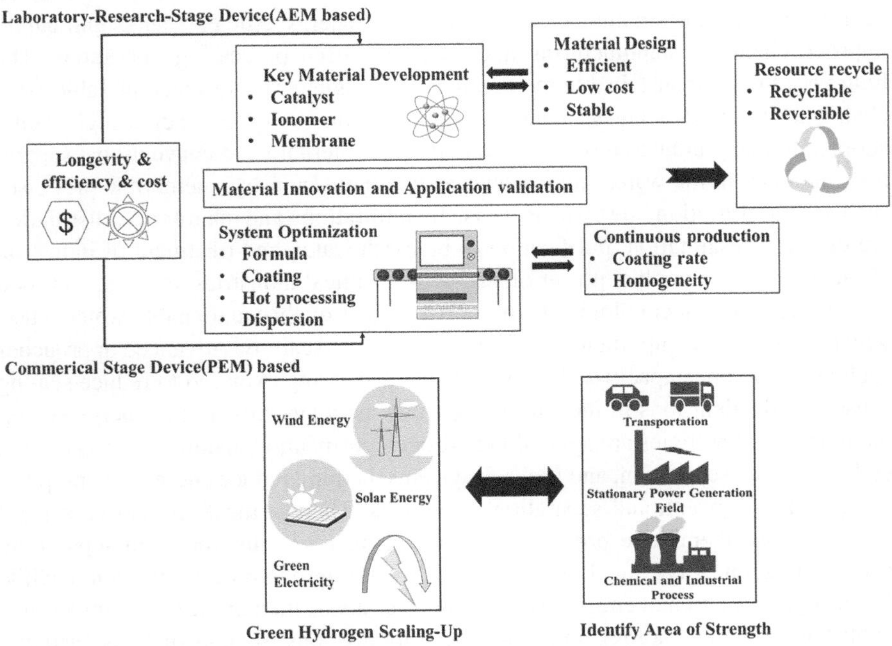


**FIGURE 10.1** Industrial ecosystem and circular economy framework for membrane technologies in electrochemical devices.

**TABLE 10.3**
**Future Roadmap for Sustainable Membrane Technologies**

| Dimension | Current Challenges | Future Roadmap |
|---|---|---|
| **Materials** | Reliance on PFSA, toxic precursors, and non-degradable waste | Shift to fluorine-free polymers, biodegradable/renewable membranes (cellulose, chitosan, lignin) |
| **Manufacturing** | High solvent use (NMP, DMF), VOC emissions, and energy-intensive drying | Green solvents, solvent-free processing, roll-to-roll scalable fabrication, low-energy curing |
| **Performance** | Trade-off between cost and durability (e.g., Nafion vs. hydrocarbon) | Composite/functionalized membranes combining low cost, high ionic conductivity, and chemical stability |
| **End-of-Life** | Landfilling and incineration of spent membranes | Closed-loop recycling, reprocessable polymers, and solvent recovery systems |
| **Policy & Regulation** | PFAS bans, REACH restrictions, weak recycling mandates | Stronger incentives for recycling, carbon pricing, clean-tech funding, and mandatory LCA reporting |

energy production, depletion of natural resources, artificial hybrid organs, neurosciences, and technical textiles. The development of innovative processes and advanced equipment that are smaller, cleaner, and more energy-efficient can significantly improve chemical manufacturing and other industrial processing operations. The future roadmap for sustainable membrane technologies was presented in Table 10.3. Owing to these advantages, membrane-based technologies are increasingly being adopted in many industrial sectors as an effective alternative to conventional separation methods. In the water and wastewater treatment industry, membrane processes such as ultrafiltration, reverse osmosis, nanofiltration, and membrane distillation are used for desalination, purification of brackish water, and treatment of industrial effluents from sectors like power plants, petrochemical industries, textiles, and food processing. These technologies help improve water quality and enable water reuse, which is especially important in regions facing water scarcity. Advanced approaches such as membrane capacitive deionization are also being explored to reduce scaling caused by divalent ions in industrial water treatment systems. In the energy sector, membrane technologies are applied in hydrogen purification, natural gas processing, carbon dioxide separation, and fuel cell systems, helping reduce energy consumption compared to conventional separation techniques. The chemical and petrochemical industries use membrane processes for solvent recovery, hydrocarbon separation, and purification of chemical products, which can improve process efficiency while lowering environmental emissions. Beyond these areas, membrane technologies also contribute to biotechnology and biomedical applications, such as dialysis, artificial organs, and purification of biological products, as well as in technical textile and advanced material manufacturing, where selective filtration and protective membrane materials are required. Overall, the growing use of membrane technologies across industries supports cleaner production processes, better resource utilization, and more sustainable industrial development.

## 10.8 CONCLUSION

Membrane technologies have gained increasing attention as an effective approach to address growing energy demands and environmental sustainability challenges. Due to their compact structure and energy-efficient separation capabilities, they are widely applied in various industrial processes. At the same time, the growing interest in environmentally friendly materials has encouraged the use of natural biopolymers as sustainable alternatives to synthetic polymers. Recent advancements in membrane fabrication also emphasize circular economy principles and adoption of green chemistry, including the use of eco-friendly solvents such as γ-valerolactone and Cyrene and renewable polymer sources. In addition, some circular strategies, such as upcycling, recycling, and membrane reprocessing, help increase membrane lifetimes and improve resource efficiency. However, the large-scale implementation of these technologies depends strongly on the sustainable availability and management of CRMs. Recent progress in hydrocarbon-based membranes, particularly sulfonated polyether ketone (SPEEK) membranes, demonstrates promising performance and cost advantages for flow batteries. Although these developments show significant progress, challenges related to durability, scalability, and economic feasibility remain unsolved. Continued research and technological innovation will therefore be essential to further improve membrane materials and support the broader adoption of sustainable membrane technologies in future energy systems.

## REFERENCES


[1] H. Prifti, A. Parasuraman, S. Winardi, T. M. Lim, and M. Skyllas-Kazacos, *Membranes for Redox Flow Battery Applications*, MDPI AG. 2012, doi: 10.3390/membranes2020275

[2] E. Panieri, K. Baralic, D. Djukic-Cosic, A. B. Djordjevic, and L. Saso, *PFAS Molecules: A Major Concern for the Human Health and the Environment*, MDPI, 1 Feb. 2022. doi: 10.3390/toxics10020044

[3] K. Cerdan et al., "Sustainability of self-healing polymers: A holistic perspective towards circularity in polymer networks," *Prog. Polym. Sci.*, vol. 152, p. 101816, 2024, doi: 10.1016/j.progpolymsci.2024.101816

[4] A. B. Yaroslavtsev, I. A. Stenina, and D. V. Golubenko, "Membrane materials for energy production and storage," *Pure Appl. Chem.*, vol. 92, no. 7, pp. 1147–1157, Jul. 2020, doi: 10.1515/pac-2019-1208

[5] L. Shi, J. Chen, S. Zhao, L. Du, and S. Ye, *Proton-exchange Membrane Water Electrolysis: From Fundamental Study to Industrial Application*, Cell Press, 21 Sep. 2023. doi: 10.1016/j.checat.2023.100734

[6] W. Song et al., "Upscaled production of an ultramicroporous anion-exchange membrane enables long-term operation in electrochemical energy devices," *Nat. Commun.*, vol. 14, no. 1, Dec. 2023, doi: 10.1038/s41467-023-38350-7

[7] P. Yadav, N. Ismail, M. Essalhi, M. Tysklind, D. Athanassiadis, and N. Tavajohi, "Assessment of the environmental impact of polymeric membrane production," *J. Memb. Sci.*, vol. 622, Mar. 2021, doi: 10.1016/j.memsci.2020.118987

[8] M. Mahnoor, R. Chandio, A. Inam, and I. U. Ahad, *Critical and Strategic Raw Materials for Energy Storage Devices*, Multidisciplinary Digital Publishing Institute (MDPI), 1 Apr. 2025. doi: 10.3390/batteries11040163

[9] H. Bardt, "Raw materials in the field of electrochemical energy storage - A risk analysis," in *AIP Conference Proceedings*, American Institute of Physics Inc., Aug. 2016. doi: 10.1063/1.4961894

[10] E. Borri, G. Zsembinszki, and L. F. Cabeza, "Evaluation of the social impact of an energy system for residential heating applications based on a novel seasonal thermal energy storage," *J. Energy Storage*, vol. 86, p. 111210, 2024, doi: 10.1016/j.est.2024.111210

[11] C. Y. Loh, A. D. Burrows, and M. Xie, *Sustainable Polymeric Membranes: Green Chemistry and Circular Economy Approaches*, American Chemical Society, 8 Aug. 2025, doi: 10.1021/acsestengg.5c00282

[12] T. M. Joseph, A. B. Unni, K. S. Joshy, D. Kar Mahapatra, J. Haponiuk, and S. Thomas, *Emerging Bio-based Polymers from Lab to Market: Current Strategies, Market Dynamics and Research Trends*, Multidisciplinary Digital Publishing Institute (MDPI), 1 Mar. 2023, doi: 10.3390/c9010030

[13] A. M. Moschovi, E. Zagoraiou, E. Polyzou, and I. Yakoumis, "*Recycling of critical raw materials from hydrogen chemical storage stacks (PEMWE), Membrane Electrode Assemblies (MEA) and electrocatalysts*," in *IOP Conference Series: Materials Science and Engineering*, IOP Publishing Ltd, Jan. 2021. doi: 10.1088/1757-899X/1024/1/012008

[14] G. Reimonn and J. Kamcev, *Techno-economic Perspective on the Limitations and Prospects of Ion-Exchange Membrane Technologies*, Elsevier Ltd., 01 Mar. 2025, doi: 10.1016/j.coche.2024.101077

[15] L. V. Garcia, Y. C. Ho, M. M. Myo Thant, D. S. Han, and J. W. Lim, *Lithium in a Sustainable Circular Economy: A Comprehensive Review*, Multidisciplinary Digital Publishing Institute (MDPI), 1 Feb. 2023, doi: 10.3390/pr11020418

# Index

Pages in *italics* refer to figures and pages in **bold** refer to tables.

For Product Safety Concerns and Information please contact our
EU representative GPSR@taylorandfrancis.com Taylor & Francis
Verlag GmbH, Kaufingerstraße 24, 80331 München, Germany